Ambient Noise in the Sea

ROBERT J. URICK

Ambient Noise in the Sea

Robert J. Urick

Adjunct Professor
The Catholic University of America
Washington, D.C.

P.O. BOX 867 • LOS ALTOS, CALIFORNIA 94023

Ambient Noise in the Sea

Revised and Edited from 1984 edition published by Undersea Warfare Technology Office, Naval Sea Systems Command, Department of the Navy, Washington, D.C.

Copyright © 1986 by Peninsula Publishing. All Rights Reserved. Printed in the United States of America. This book or parts thereof, may not be reproduced in any form without permission of the publishers.

Library of Congress Catalog Card Number: 85-063685
ISBN 0-932146-13-9

PREFACE

Noise is unwanted sound. Ambient noise is the noise that is the typical or persistent noise background at some spot that is independent of the means used to observe it. Ruled out are all forms of "self-noise" caused by the hydrophone and its platform, as well as occasional transitory sounds like that of a passing ship or an earthquake. Wanted sounds are called <u>signals</u> and are often similar to the unwanted sounds called noise.

The literature of ambient noise in the sea is a prolific one. Next to propagation, it has the largest literature of all aspects of underwater acoustics, partly (one may jocularly surmise) because of the relative ease by which it can be observed and measured. Until World War II nothing was known about it in a scientific way; now, some forty years later, a complete bibliography would include several thousand reports and papers, many of which are, sad to say, classified.

The purpose of the present addition to this already abundant literature is to summarize the main features of the subject and to give the reader an entry into the unclassified literature on some topic that may be of interest. What follows contains little tutorial matter, it being assumed that the reader has some appreciation of jargon like "spectrum level" and "dB re 1μPa." Readers who have not may be referred to text books on acoustics, and specifically to <u>Mechanics of Underwater Noise</u> by D. Ross or to <u>Principles of Underwater Sound</u> by myself.

R. J. Urick
March 1984

ACKNOWLEDGEMENTS

My gratitude is extended to B. G Hurdle of the Naval Research Laboratory for making documents available and for helpful discussions, to C. D. Smith and A. P. Franceschetti of the Undersea Technology Office, Naval Sea Systems Command, for their encouragement and financial support, and to R. Wagstaff of the Naval Ocean Research and Development Activity, J. Northrop of the Naval Ocean Systems Center, W. R. Schumacher and D. Browning of the Naval Underwater Systems Center, and D. F. Young of Mandex, Inc. for their careful review of the manuscript. Finally, as Adjunct Professor at The Catholic University of America, my thanks are due to the staff and personnel of the School of Engineering and Architecture for their assistance in so many ways in the preparation of this report.

TABLE OF CONTENTS

	Page
1. INTRODUCTION.	1-1
1.1 Historical Summary	1-1
1.2 Existing Bibliographies and Summaries.	1-3
1.3 Scope and Purpose of This Survey	1-4
2. SOURCES OF AMBIENT NOISE.	2-1
2.1 Introduction	2-1
2.2 Ultra-Low Band (< 1 Hz).	2-1
2.3 Infrasonic Band (1 to 20 Hz)	2-4
2.4 Low Sonic Band (20 to 200 Hz).	2-13
2.5 High Sonic Band (200 to 50,000 Hz)	2-18
2.6 Ultrasonic Band (> 50 kHz)	2-27
2.7 Processes of Surface Noise Generation.	2-30
2.8 Shallow Water.	2-33
2.9 Effects of Propagation on Noise.	2-37
2.10 Summary.	2-42
3. VARIABILITY OF AMBIENT NOISE.	3-1
3.1 Introduction	3-1
3.2 Noise Variability in Deep Water.	3-3
3.2.1 Short Period Variations	3-3
3.2.2 Medium Period Variations.	3-3
3.2.3 Long Period Variations.	3-9
3.3 Noise Variability in Shallow Water	3-12
3.4 Correlation Between Frequency Bands and the Effect of Beamwidth.	3-18
3.5 Summary.	3-21
4. DEPENDENCE ON RECEIVER DEPTH.	4-1
4.1 Introduction	4-1
4.2 Theory	4-1
4.2.1 Surface Noise	4-1
4.2.2 Shipping Noise.	4-2
4.3 Field Observations	4-4
4.4 Noise in the Surface Duct.	4-12
4.5 Effect of Bathymetry	4-12
4.6 Summary.	4-14

TABLE OF CONTENTS (Continued)

		Page
5.	DIRECTIONALITY OF AMBIENT NOISE	5-1
	5.1 Introduction	5-1
	5.2 Theoretical Considerations	5-1
	5.3 Vertical Directionality	5-4
	5.4 Variation of Vertical Directionality with Depth	5-13
	5.5 Horizontal Directionality	5-15
	5.6 Summary	5-22
6.	COHERENCE OF AMBIENT NOISE	6-1
	6.1 Introduction	6-1
	6.2 Theory	6-2
	6.3 Field Observations	6-8
	6.4 Summary	6-14
7.	INTERMITTENT SOURCES OF NOISE	7-1
	7.1 Introduction	7-1
	7.2 Biologic Noises	7-1
	7.2.1 Whales, Dolphins and Porpoises	7-2
	7.2.2 Croakers	7-6
	7.2.3 Snapping Shrimp	7-9
	7.3 Twenty-Cycle Pulses	7-11
	7.4 Non-Biologic Noises	7-14
	7.4.1 Rain	7-14
	7.4.2 Earthquakes, Explosions and Volcanos	7-17
	7.4.3 Surf	7-18
	7.5 Summary	7-19
8.	NOISE IN THE ARCTIC	8-1
	8.1 Introduction	8-1
	8.2 Sources of Noise in the Arctic	8-1
	8.2.1 Ice Cracking	8-1
	8.2.2 The Wind	8-2
	8.2.3 Moving Ice Masses	8-7
	8.3 The Marginal Ice Zone	8-12
	8.4 Biological Noise	8-14
	8.5 Low-Frequency Noises	8-14
	8.6 Summary and Compilation	8-15

TABLE OF CONTENTS (Continued)

	Page
9. STATE OF PRESENT KNOWLEDGE.	9-1
APPENDIX - IDEALIZED AMBIENT NOISE SPECTRA BASED ON THE LITERATURE	A-1
LIST OF REFERENCES.	R-1
INDEX	I-1

LIST OF FIGURES

Figure No.		Page
2-1	Low frequency output of a hydrophone showing tidal changes caused by a combination of pressure and temperature changes .	2-3
2-2	World War II low frequency ambient noise measurements, made at various U.S. and U.K. locations, in various band widths, mostly from 2 to 20 Hz, in shallow water 12 to 110 ft deep. .	2-5
2-3	A compilation of infrasonic ambient noise levels.	2-6
2-4	Noise spectrum in the range 5 to 30 Hz from bottomed hydrophone measurements in deep water (2400 m) in an area of dense ship traffic.	2-7
2-5	Noise at 12.5 Hz at three sites distant from a shipping lane and at one site near a major shipping lane	2-9
2-6	Noise difference between the North and South Pacific.	2-10
2-7	Theoretical spectra of current-induced pseudo-noise	2-12
2-8	Effect of wind speed at different frequencies	2-15
2-9	Noise levels at four stations in the Mediterranean Sea, averaged over three days at each station.	2-17
2-10	The "Knudsen Curves".	2-20
2-11	Noise spectrum level versus wind speed at different frequencies in the northwest Atlantic	2-21
2-12	Variation of level with wind speed at various frequencies north of St. Croix.	2-22
2-13	Noise spectra measured between Australia and New Zealand. . .	2-24
2-14	Spectrum levels at two depths in the Parece Vela Basin west of Guam.	2-25
2-15	Surface noise and molecular noise in the band from 10 kHz to 1000 kHz	2-29
2-16	Theoretically computed spectra for the two principal sources of wind-induced low-frequency noise	2-31

vi

LIST OF FIGURES (Continued)

Figure No.		Page
2-17	Shallow water ambient noise spectra, with Beaufort wind number as a parameter	2-35
2-18	Noise spectra at six locations in the Coral, Solomon and Bismark Seas.	2-36
2-19	Difference in noise level between a shallow (30 fms) and a deep hydrophone (900 fms) 3.3 miles away off Bermuda.	2-38
2-20	Cross-section of the continental shelf and slope off Pt. Arena, California, showing how a ray leaving the source at an angle of 15° becomes a ray in the deep sound channel	2-40
2-21	Conceptual view of noise spectra over the range 1 to 10,000 Hz, showing shipping noise plus two noise processes at low and high frequencies	2-43
3-1	Time scales for variability of four noise samples	3-2
3-2	Cumulative distribution curves of 114 1-1/2-minute samples taken at hourly intervals over a 2-1/2-day period at five depths north of St. Croix	3-5
3-3	Sound spectrograms at three depths obtained hourly north of St. Croix	3-6
3-4	Variability of noise profiles from hour to hour	3-8
3-5	Monthly average ambient noise levels at two locations near Bermuda	3-11
3-6	Playouts of 1-minute noise samples taken at hourly intervals at two locations off the east coast of the U.S.	3-13
3-7	1000 Hz noise levels at various shallow water locations obtained from the literature.	3-14
3-8	Correlation coefficients of 10.7-minute samples of noise between different 6 Hz frequency bands.	3-19
3-9	Relative fluctuation spectra.	3-20
4-1	Theoretical attenuation of surface-generated noise with depth caused by absorption, assuming straight-line paths, an infinite water depth and a constant absorption coefficient with depth.	4-3

LIST OF FIGURES (Continued)

Figure No.		Page
4-2	Noise profiles at bottomed hydrophones off Bermuda.	4-6
4-3	Hydrophone string and velocity profile for measurements of the depth profile north of St. Croix	4-7
4-4	Depth profile of noise north of St. Croix measured in 1/3 octave frequency bands.	4-9
4-5	Velocity profile, hydrophone depths and depth profiles measured in the northeast Pacific Ocean	4-10
4-6	Depth profiles at and below the Deep Sound Channel axis in the northeast Pacific Ocean	4-11
4-7	Ambient noise spectra within and below a surface duct	4-13
5-1	Geometry of a simple model for the vertical directionality of noise from random sources on a plane surface, each radiating like $\cos^n \theta$.	5-2
5-2	Vertical directional patterns at three sea states in the band from 750 to 1500 Hz as obtained by Becken (122) with a 3-dimensional array	5-5
5-3	Vertical directional patterns as obtained by Forster in a 1/2-octave band centered at 2000 Hz	5-7
5-4	Vertical directional patterns in 1/3-octave bands at four frequencies in sea state 3.	5-8
5-5	Ambient noise directional patterns at 112 and 891 Hz as reported by Axelrod, Schoomer and Von Winkle (124).	5-10
5-6	Directional patterns of ambient noise at two frequencies.	5-11
5-7	Ambient noise directionality measurements at two frequencies in the South Fiji Basin	5-12
5-8	Vertical noise patterns on the Sound Channel axis (1200 m) in the South Fiji Basin	5-14
5-9	Vertical directional patterns at two depths at a frequency of 38 Hz.	5-16
5-10	Horizontal patterns at two vertical angles.	5-17

viii

LIST OF FIGURES (Continued)

Figure No.		Page
5-11	A hypothetical example of the directionality of shipping noise as it would be observed with an array of infinite resolution.	5-19
5-12	Horizontal directionality of ambient noise.	5-21
6-1	Expressions for the coherence of unidirectional and isotropic noise for a single frequency and for a band of frequencies, with and without an electrical time delay	6-4
6-2	Coherence functions vs. normalized spacing.	6-5
6-3	Correlogram contour plots showing the correlation coefficient as a function of normalized spacing and time delay.	6-7
6-4	The Trident Vertical Array at Bermuda, with which a number of coherence measurements by different investigators have been obtained	6-9
6-5	Octave band low frequency coherence of ambient noise as found with the Trident Vertical Array, normalized to the mid-band wavelength λm.	6-11
6-6	Time-delay correlograms at a high and a low wind speed in the 200-400 Hz octave band for various vertical hydrophone separations	6-12
6-7	Regions of dominance of ship and wind noise as deduced from the types of correlograms obtained.	6-13
6-8	Correlograms at different vertical separations in a surface duct, showing two types of noise.	6-15
7-1	Mean level at 1 meter, in a 120 Hz band, of the whistles and clicks from herds of the common dolphin <u>Delphinus delphis</u>	7-4
7-2	Sound spectrogram of a humpback whale	7-5
7-3	Summary of the components of ambient noise as observed in the Timor Sea north of Australia.	7-8
7-4	The noise of snapping shrimp.	7-10
7-5	Sound spectrogram of long twenty-cycle pulses as recorded off San Clemente Island, California	7-13

ix

LIST OF FIGURES (Continued)

Figure No.		Page
7-6	Theoretical rain noise spectra of Franz compared with two field measurements.	7-16
8-1	Diurnal variations in air temperature in degrees C and the noise level in the 150 to 300 Hz band	8-3
8-2	Ice-cracking pulses and their envelopes, frequency band 200 to 600 Hz	8-4
8-3	Spectra of ice-cracking noise during the nighttime hours.	8-5
8-4	Cumulative distribution curves of the amplitude of (A) ice-cracking sounds in shore-fast springtime ice, and (B) sounds made by rafting summer ice, as observed in the Canadian Archipelago.	8-6
8-5	Spectra of wind noise under shore-fast ice for several wind speeds	8-8
8-6	Wind noise levels under a continuous ice cover during the month of February in the frequency band 3.2 to 6.4 kHz.	8-9
8-7	Noise level vs. wind speed at 141 Hz.	8-11
8-8	Noise level on either side of a sharp ice edge.	8-13
8-9	Sound spectrograms showing the low frequency buoyant bobbing frequencies of late summer ice floes.	8-16
8-10	A compilation of noise spectra in the Arctic.	8-17
A	Spectra of ambient noise in the band 1 to 1000 Hz, based upon measurements in the literature and present models for noise spectra in various frequency ranges	A-2

LIST OF TABLES

Table No.		Page
1-1	Beaufort Wind Force and Sea State Numbers Vs Wind Speed	1-5
2-1	Empirical Coefficients for Wind-Dependent Noise	2-26

1. INTRODUCTION

By ambient noise we mean the prevailing, sustained unwanted background of sound at some spot in the ocean. It excludes momentary, occasional sounds, such as the noise of a close-by passage of a ship or of an occasional rain squall. It is the background of noise, typical of the location and depth where a measuring hydrophone is located, against which a "signal," such as the sound of a submarine or the echo from a target, must be detected. Ambient noise also excludes all forms of self-noise, such as the noise of current flow around the measurement hydrophone and its supporting structure, and obviously must exclude all forms of electrical noise. Thus, ambient noise is what is left over, so to speak, after identifiable, occasional noise sources are accounted for.

1.1 Historical Summary

Strange to say, the ambient background of noise in the sea was ignored during the years prior to World War II, when fairly sophisticated echo-ranging sonars were being developed and installed on ASW vessels. The reasons for this lack of attention are several-fold. First, both the number of engineers and scientists working in sonars, as well as the level of funding was extremely small by post-war standards. Secondly, attention was directed almost exclusively toward echo-ranging sonars, in which the ambient background is apt to be neligible compared to the background of reverberation and self-noise. Thirdly, no absolute measurements could be made at that time because standard hydrophones and calibration techniques had not yet become available.

It was only during World War II when manpower, money and a realization of the value of research, came into being. Also, an added practical incentive for investigation of noise in the sea was the emergence of the acoustic mine, in which the level of the ambient background must be known in order to establish the sensitivity requirements for the firing mechanism. Early in the war a research group was established at San Diego as part of the National Defense Research Committee Division 6, Section 6.1, under the direction of V.O. Knudsen. This group made ambient noise measurements in a number of bays, harbors and off-shore areas and in 1944 wrote a comprehensive report (1)* that was later summarized in a now-classic paper in the Journal of Marine Research dated 1948. Later, in 1954, the present author, along with Aubrey Pryce, summarized ambient noise as part of a more comprehensive report (2) intended for practical use by engineers and analysts.

Theoretical and field work on ambient noise continued at a low level until about 1960, when there was a sudden surge of research activity. The present list of references shows a total of 19 references in the fifteen year period 1945-1960, whereas there are 131 references in the fifteen years between 1961 and 1976. This renewed activity, which has continued unabated up to the present time, was brought about by the somewhat belated interest of the Navy in passive sonars that has led to the development of systems such as SOSUS, towed line arrays, and submarine sonars for long range detection and surveillance. This more recent work has provided data and understanding for frequencies below 100 Hz and has resulted in a knowledge of the causes, or sources, of noise over the entire frequency range of sonar interest.

---

*Numbers in parentheses refer to items in the list of references (pp. R-1 to R-14).

1.2 Existing Bibliographies and Summaries

The literature on ambient noise is enormous, partly because ambient noise is relatively easy to measure and study. Indeed, all that is needed to make a primitive measurement of the spectrum level and shape for the noise in some bay, harbor or inshore area of particular interest is a calibrated hydrophone suspended from a rowboat.

Except for the subject of propagation, ambient noise now has the largest literature of all aspects of underwater sound. To cope with this relatively vast literature, a number of bibliographies, or reference lists, have been compiled. All are dated between 1973 and 1975. Perhaps the earliest is one by Wagstaff (3) having 650 references divided into sub-topics. Another is by Gold and Columbo (4) containing 1100 references, still admittedly incomplete. In order to make this literature more useful, a data bank was established at this time. A report by Ingalsbee (5) lists some 1450 references of items in the data bank as of 1975, superceding a partial list of some 1300 by Palumbo and Gold (6). In another report by Gold and Palumbo (7), the number of reports and papers existing in the data bank is given by area and topic. For example, there are 260 reports on the ambient noise levels in the North Atlantic Ocean, but there are none for the South Atlantic.

Based on this literature, a number of summaries of information on ambient noise in the sea are extant. The classic summary is a paper by Wenz (8), published in 1962, which discusses the sources of noise in the sea in a comprehensive way and presents some new data; some additional details are included in a short subsequent paper (9) dated 1963. Also, in order to bring the Wenz report up to date, a review paper dated 1965 by Arase and Arase (10), containing 76 references, has been published. The most recent summary is one by Crouch (11), dated 1972, a report that is excellent for its time, on the subjects of directionality and coherence. Wenz (12) also reviewed the subject about this time. Finally, we should not fail to mention a few Russian summary papers. One by Furduev (13), dated 1964, reviews the subject of noise due to water motion; however, only 4 of the 36 cited references are of Russian origin. The most recent Russian review paper, dated 1975, is a translation of a paper by Furduev (14), and is based on 125 cited papers that do contain a sizeable number of Russian references.

1.3 Scope and Purpose of This Survey

The present survey attempts to bring these existing summaries up to date. Like them, it is incomplete and is based on literature references that were (1) available, (2) significant in content, and (3) unclassified. It is intended as a guide to the many aspects of noise in the sea. Included are compilations of the levels of noise in shallow water at 1 kHz (Figure 3-7), the noise under an ice cover (Figure 8-10), and some generalized deep-water spectra based on measured data (Appendix). In the figures, the spectrum levels have been converted, where necessary, from the original pressure references of 1 dyne/cm^2 or 0.0002 dyne/cm^2 to the present reference standard of 1 micropascal (μPa).

Ambient noise spectra in the literature usually have sea state, Beaufort number, or wind speed as a parameter. Table 1-1 gives the relationship between these quantities.

Table 1-1. Beaufort Wind Force and Sea State Numbers Vs Wind Speed
Reference 15

Beaufort Number	Sea State	Wind Speed Knots	Wind Speed Meters/Sec
0	0	<1	0 - 0.2
1	1/2	1 - 3	0.3 - 1.5
2	1	4 - 6	1.6 - 3.3
3	2	7 - 10	3.4 - 5.4
4	3	11 - 16	5.5 - 7.9
5	4	17 - 21	8.0 - 10.7
6	5	22 - 27	10.8 - 13.8
7	6	28 - 33	13.9 - 17.1
8	6	34 - 40	17.2 - 20.7

2. SOURCES OF AMBIENT NOISE

2.1 Introduction

If you were to take a very broad band calibrated hydrophone, place it on the sea bed (for stability), and plot the spectrum of the observed background over a wide frequency range, you would find a number of peculiarities in the shape and slope of the spectrum. There would be some frequency bands where tonal components occur. There would be others where the spectrum would be continuous and negatively sloping ("pink" noise), separated by portions where the spectrum would be flat ("white" noise) or even reversed in slope. From this observation you would conclude, correctly, that different sources of noise must exist and be prevalent in different regions of the spectrum.

In the following, these various and diverse sources will be discussed. The overall frequency range, extending over some seven decades of frequency, will be divided somewhat arbitrarily into five frequency bands, in each of which the prevailing sources of noise appear to be different.

2.2 Ultra-Low Band (< 1 Hz)

Almost nothing is known about the noise in this band; except for one series of measurements, all that we can do is to speculate and to surmise. The measurements just referred to were made long ago, and were only recently published by Nichols (16). The measurements were reported in octave bands and were made with bottomed hydrophones at depths of 13, 300 and 1200 m off the island of Eleuthera in the Bahamas; the averaged 300-m and 1200-m results are included with measurements in the next higher band in Figure 2-3. More recent measurements were made in 1978-1979, also off Eleuthera, by Talpey and Worley (220) and were analyzed in 0.1 Hz bands over the range 0.1-12.5 Hz. A summary of the data obtained on 5 days is the shaded area of Fig. 2-3 for a depth of 3500 m. During the observation period the wind speed varied from 0-20 knots; a correlation with wind speed was noted, with a correlation coefficient of 0.8 between 2 and 3 Hz, falling off above and below this band. Strangely, no tonal components in the noise were reported, not even in the region below 1 Hz. In this frequency band below 1 Hz, it is certain that the spectrum, as observed using a pressure-sensitive hydrophone, would be dominated by line components. These tonal components would not necessarily represent acoustic pressures propagating with the velocity of sound, but would more likely be of hydrostatic or hydrodynamic origin. An example would be the complex ensemble of discrete frequencies due to the tides, with lunar and solar periods and their harmonics. In addition, the pressures due to waves and swell, particularly the latter, would be likely to be significant only at depths less than about a wavelength of the waves, except for the non-linear wave interactions to be discussed later on.

It is interesting to note that the motion of the sea surface produces sound in the air above it as well as in the water below. Waves in the air, with periods near 5 sec, have been observed and attributed to this cause and have been called "microbaroms" [17][18], in analogy with the much-studied microseisms traveling through the earth.

That the motion of the sea bottom can give rise to pressures in the water above was pointed out by Urick [19], who found reasonable agreement with acoustic data at frequencies above 1 Hz with the levels computed from independent seismic measurements of bottom motion. In this connection, the agreement of the measurements of Nichols cited above [16] in the 0.1 to 1 Hz range with data taken on both a hydrophone and a seismometer by Schneider and Backus [20] is particularly striking.

A combination of tidal pressures and associated temperature changes--all hydrophones are to some degree temperature sensitive due to the "pyroelectric" effect--can produce enormous outputs from a hydrophone. Figure 2-1 shows the observed voltage output due to these causes, amounting to almost 10 volts at the terminals of the hydrophone. The pyroelectric effect alone was said to amount to 12 volts per degree Celsius. Such effects, together with the pressures or "pseudo-noise" due to water motion (see below) make valid

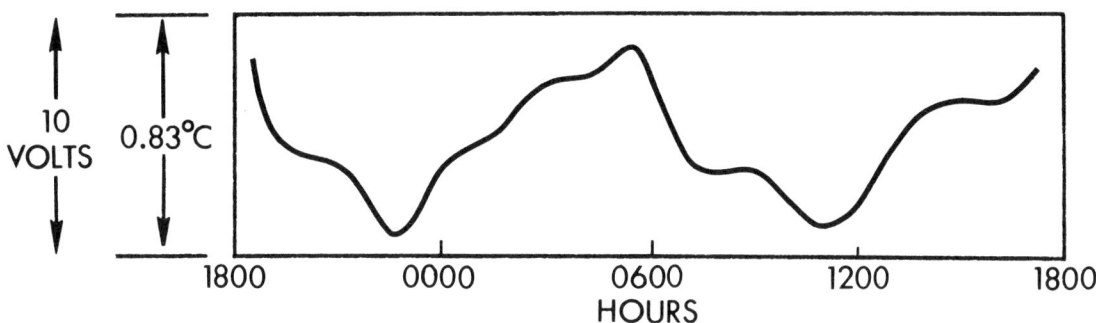

Figure 2-1. Low frequency output of a hydrophone showing tidal changes caused by a combination of pressure and temperature changes. The temperature sensitivity of the hydrophone was 21.6 dB re 1 volt per degree C, or about 12 volts per degree. Reference 219.

acoustic observations difficult in this neglected band of the overall ambient noise spectrum.

2.3 Infrasonic Band (1 to 20 Hz)

This band has received considerable recent attention. It contains the strong blade-rate fundamental frequency of propeller-driven vessels, plus one or two of its harmonics, and the band is therefore of great interest to low frequency passive sonars.

During World War II, data in this band was obtained in connection with the then-new acoustic mine, in order to provide the maximum sensitivity requirements of the mine actuation circuitry. The data obtained in these early days fell in the cross hatched area of Figure 2-2. This range of data, while 20 dB wide, is on the whole higher than more modern measurements.

In the lower part of the Infrasonic Band, in the range 1 to 5 Hz, the measurements in deep water show a steep spectral slope of about 10 dB per octave, as shown by the compilation of Figure 2-3. In the frequency region 5 to 20 Hz, in ocean areas where ship noise begins to be strong, the measured spectrum levels begin to show a reverse slope, suggesting the emergence of shipping noise at the upper end of this frequency band. An example is the spectrum reported by McGrath (21) shown in Figure 2-4, as obtained with a hydrophone located on the Mid-Atlantic Ridge at a depth of 2400 m. This noise was attributed to distant ship traffic; local ship traffic and sea quakes as recorded by nearby seismometers increased the hydrophone levels above these values for only short periods of time. On the other hand, in the absence of ship traffic, or in shallow water, the spectrum continues to fall off and to remain dependent on wind speed over the entire band.

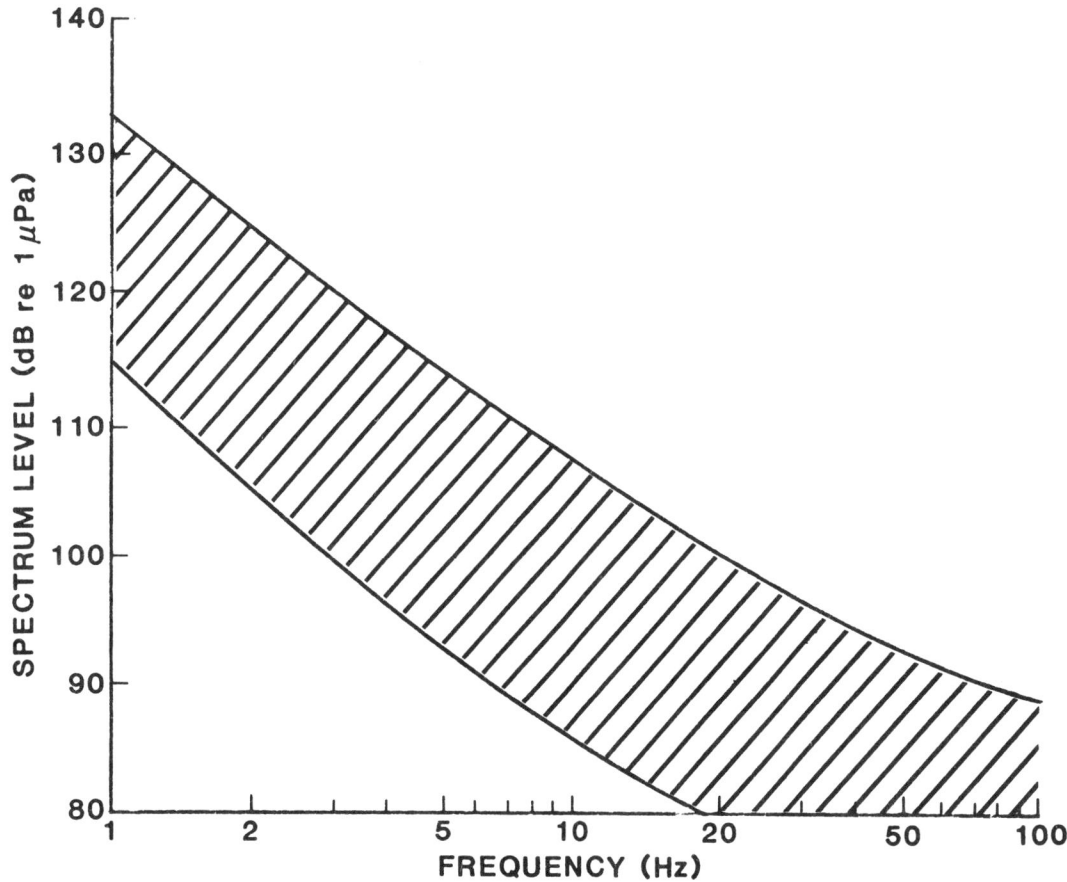

Figure 2-2. World War II low frequency ambient noise measurements, made at various U.S. and U.K. locations, in various band widths, mostly from 2 to 20 Hz, in shallow water 12 to 110 ft deep. Reference 2.

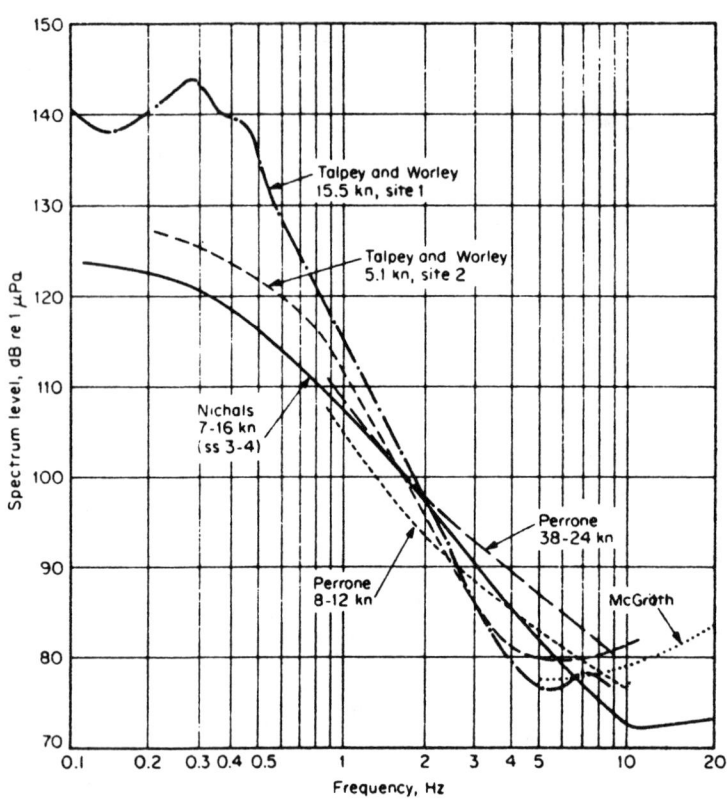

Figure 2-3. A compilation of infrasonic ambient noise levels. Solid curve: data of Nichols (16), depths 300 and 1200 m averaged, off Eleuthera, octave band analyses. Dotted curves: data of Talpey and Worley, according to Nichols (16), depth 4300 m at site 1, 3500 m at site 2, 25 miles apart, south of Bermuda, 0.0043-Hz band analyses. Dashed curve: data of Yen and Perrone (53), depth 900 m, south of Bermuda. Dots: data of McGrath (21), depth 2400 m, Mid-Atlantic Ridge, third-octave bands.

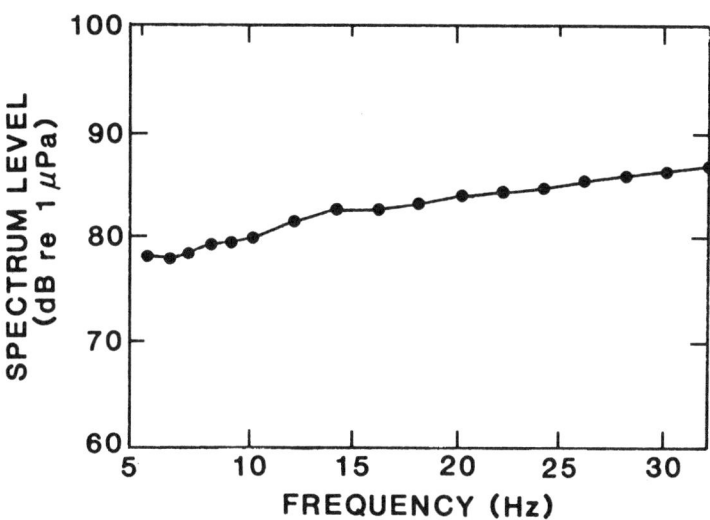

Figure 2-4. Noise spectrum in the range 5 to 30 Hz from bottomed hydrophone measurements in deep water (2400 m) in an area of dense ship traffic. Reference 21.

The importance of the place of observation on the noise in this and higher frequency bands is illustrated in Figures 2-5 and 2-6. Figure 2-5 shows data at 12.5 Hz for four locations, three of them distant from, and one close to, a shipping lane. A clear wind dependence occurs at the distant sites, but something other than the wind affects the noise near the shipping lane. Because of the scarcity of shipping, the South Pacific Ocean was found by Kibblewhite (22) to be more quiet, by about 15 dB, at frequencies less than about 200 Hz, than a location in the North Pacific. This difference in location is shown in Figure 2-6.

Perrone (23) made recordings over an 8-day period in water 614 fathoms deep on the slope of the Grand Banks. A definite wind dependence was found in the band 1 to 4 Hz, but not at higher frequencies, where the noise was dominated by fishing boats in this heavily fished area. In short, the noise in the upper part of this band, as well as in the next higher frequency band, is strongly dependent upon location, relative to the presence of ship traffic, out to considerable distance from the recording site.

A source of difficulty in making of measurements in the Infrasonic region is pseudo-noise (a name coined by Lighthill). This is a form of self-noise resulting from the presence of the hydrophone and its supporting structure in a current. There are always currents in the sea, even at great depths; the placement of a hydrophone in a current results in noise originating in a number of ways. One is by the impingement of patches of turbulent water, carried along by the current, that create variable pressures when they strike the face of the hydrophone. A similar effect is caused by temperature microstructure, which produces noise through the pyroelectric sensitivity of transducer materials. A third effect is flow-noise caused by the local boundary-layer turbulence and vortex shedding produced by the presence of the

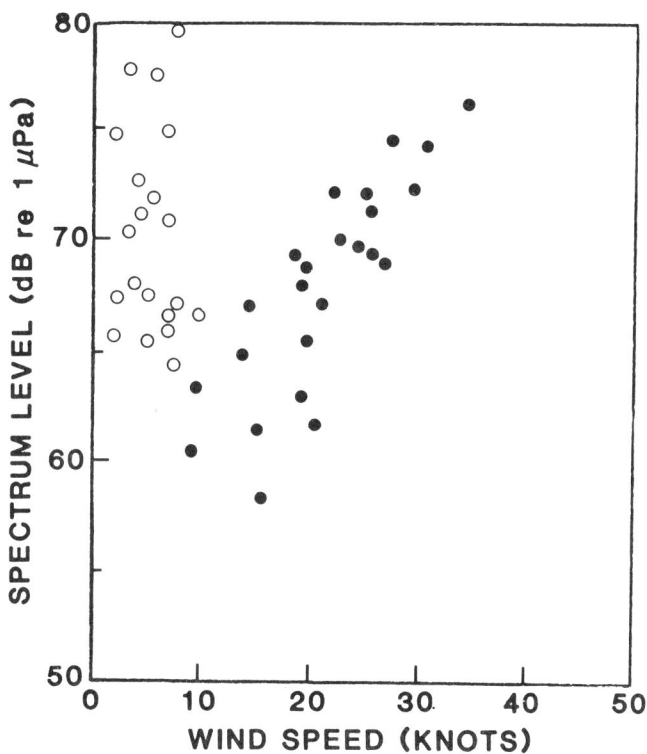

Figure 2-5. Noise at 12.5 Hz at three sites distant from a shipping lane (dots), and at one site near a major shipping lane (circles). Reference 51.

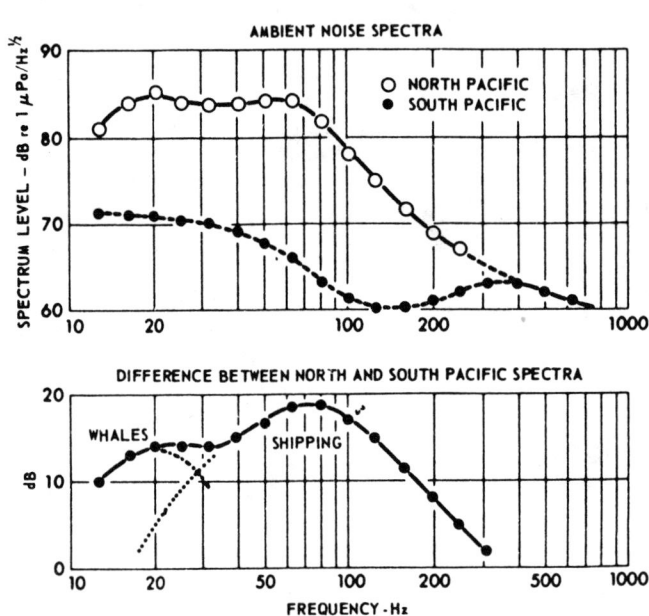

Figure 2-6. Noise difference between the North and South Pacific. Reference 22.

hydrophone and its support. <u>Cable strumming</u> falls in this category if the hydrophone is supported by a flexible cable. An evaluation of these forms of self-noise has been made by Strasberg (24). One of them--the pyroelectric effect--was estimated to be insignificant if the sensitive element is protected by a thermal insulator 1 mm or more in thickness. The other two--turbulent impingement and turbulence generated by the hydrophone itself--were evaluated, and, using reasonable estimates for the required parameters, the theoretical spectra for a 1/2- and 1-knot current were found to be roughly comparable with measured data in the region 1 to 10 Hz, as shown in Figure 2-7.

In the laboratory McGrath (25) measured the flow-noise picked up by an H-58 cylindrical hydrophone 70 mm in diameter and 0.43 m in length. This hydrophone was the same as that used at sea on the bottom on the Mid-Atlantic Ridge (see Figure 2-4). The laboratory data obtained in a tank rotating at speeds of 0.25, 0.40 and 0.45 kn were found to be about the same as the open sea measurements at frequencies between 5 and 10 Hz, but to have a reverse spectral slope. Faired and frame-mounted hydrophones gave lower levels at frequencies below 3 or 4 Hz. In Russian work in coastal waters, Bardyshev (26) placed a "flow deflector"--evidently a sphere of dense fabric 0.5 m in diameter surrounding the hydrophone--and found reductions in a 0.6 m/sec current of 24 dB in the pseudo-noise level without attenuation of a signal in the range 2 to 20 Hz. In short, great care has to be exercised in making measurements of ambient noise in the Infrasonic and Very Low Band to avoid contamination by a variety of non-acoustic noises due to water motion. Doubtless, some extant data, such as the World War II data of Figure 2-2, has been contaminated by such noise.

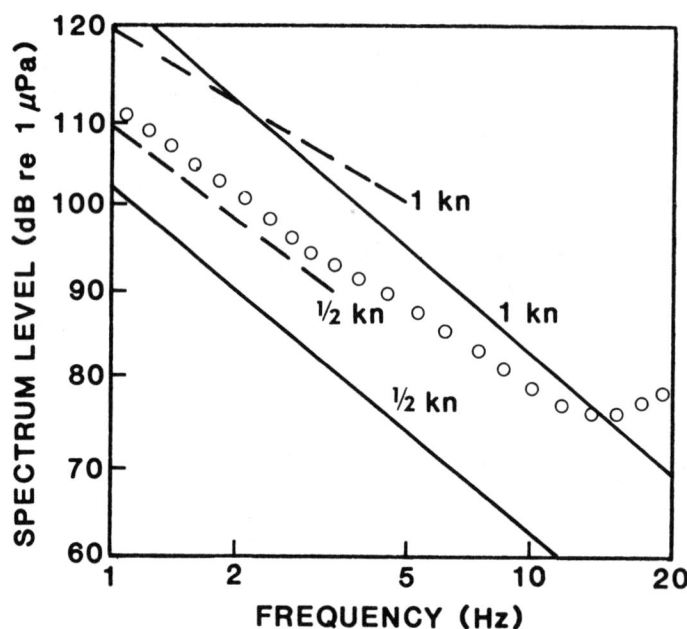

Figure 2-7. Theoretical spectra of current-induced pseudo-noise. The dashed lines show turbulence impingement. The solid lines show flow-induced turbulence (vortex shedding, local turbulence about the sensor and its mounting). The circles show measured data from Figure 2-3. Reference 24.

2-12

One technique that would be free of such contamination would be to use a neutrally buoyant drifting sensor package. Another would be to use two separated hydrophones and to correlate them together, on the expectation that true acoustic noise will be correlated if the separation distance is not too great, but the pseudo-noise will not. This method was employed by Buck and Greene (27) with hydrophones suspended from holes in the Arctic ice to a depth of 100 feet and a separation distance of 60 ft. The maximum correlation coefficient at zero time delay was 0.99 (little or no contamination) and the minimum was 0.19 (high contamination). By a simple analysis it was shown that if the measurement intensities (acoustic plus pseudo-noise) at the two hydrophones are I_1 and I_2, then the mean true acoustic intensity I_T is given by $I_T = \rho (I_1 \cdot I_2)^{1/2}$, where ρ is the cross-correlation between the outputs of the two hydrophones. In this way, it is possible to sort out the two types of noise.

Noise measurements have been reported in the band 11 to 45 Hz by Lomask and Saenger (28) in a large, deep, inland lake (Lake Pend Oreille, Idaho) with an anchored buoyed hydrophone. Under extremely calm conditions--that is, in the absence of wind, waves, shipping and marine life--the measured levels, well above electronic noise, were 10 dB below Knudsen Sea State Zero (see Figure 2-10).

2.4 Low Sonic Band (20 to 200 Hz)

This frequency band is characterized by the noise of distant shipping in areas where distant ships are prevalent. In areas remote from shipping lanes, the noise in this band continues to depend on wind speed, as it does at both lower and higher frequencies.

Figure 2-8 shows noise spectra averaged over a 27-day period during the month of January using a nondirectional hydrophone located 35 miles south of Bermuda and analyzed in 1/3-octave bands, as reported by Perrone (29). Wind-dependent noise occurs in the 11 and 17 Hz bands, as well as at frequencies above 200 Hz. In between, in the band 20 to 200 Hz, the non-wind dependent noise, with peaks at 20 and 60 Hz, is attributed to a combination of biological sources (20 cycle pulses), shore activities (60 Hz) and distant ocean ship traffic.

The importance of location in determining the noise in this frequency band is illustrated by the measurements of Cato (30) at 40 sites in the waters near Australia. A wind-speed dependence was observed at all frequencies from 22 to 5000 Hz for wind speeds above 5 m/sec. At and below this wind speed, however, the levels could be correlated with the density of ship traffic; for example, in the Tasman Sea east of Australia, the highest levels occurred at low wind speeds, as a result, it was believed, of the greater density of traffic in this area. These conclusions concerning the prevalence of noise due to the wind in the Southern Hemisphere were verified by the observations of Bannister and others (31) in waters north of New Zealand, as well as more recently by Burgess and Kewley (32) off Australia.

At the other extreme of ship traffic density, we may mention observations made in the region of heavy traffic southwest of the English Channel. Here, with a string of hydrophones extending down to 2764 m in 3956 m of water, King (33) found no depth variation and no wind dependence at any frequency from 30 to 4000 Hz. Even at frequencies as high as 4000 Hz the mean observed levels were the same as those for a wind force Beaufort 5 of the "Knudsen" curves (see Figure 2-10), even though the wind never exceeded Beaufort 4.

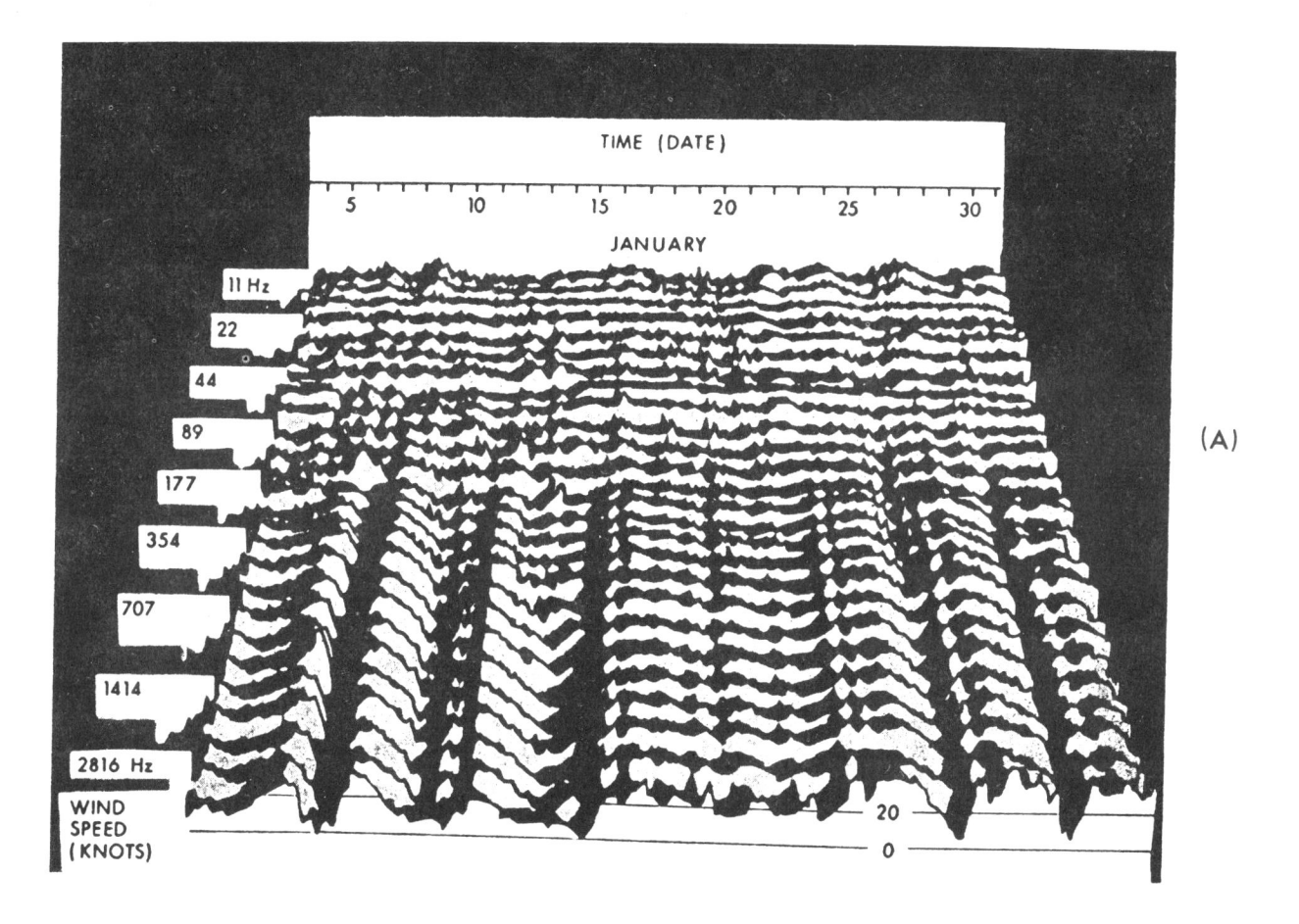

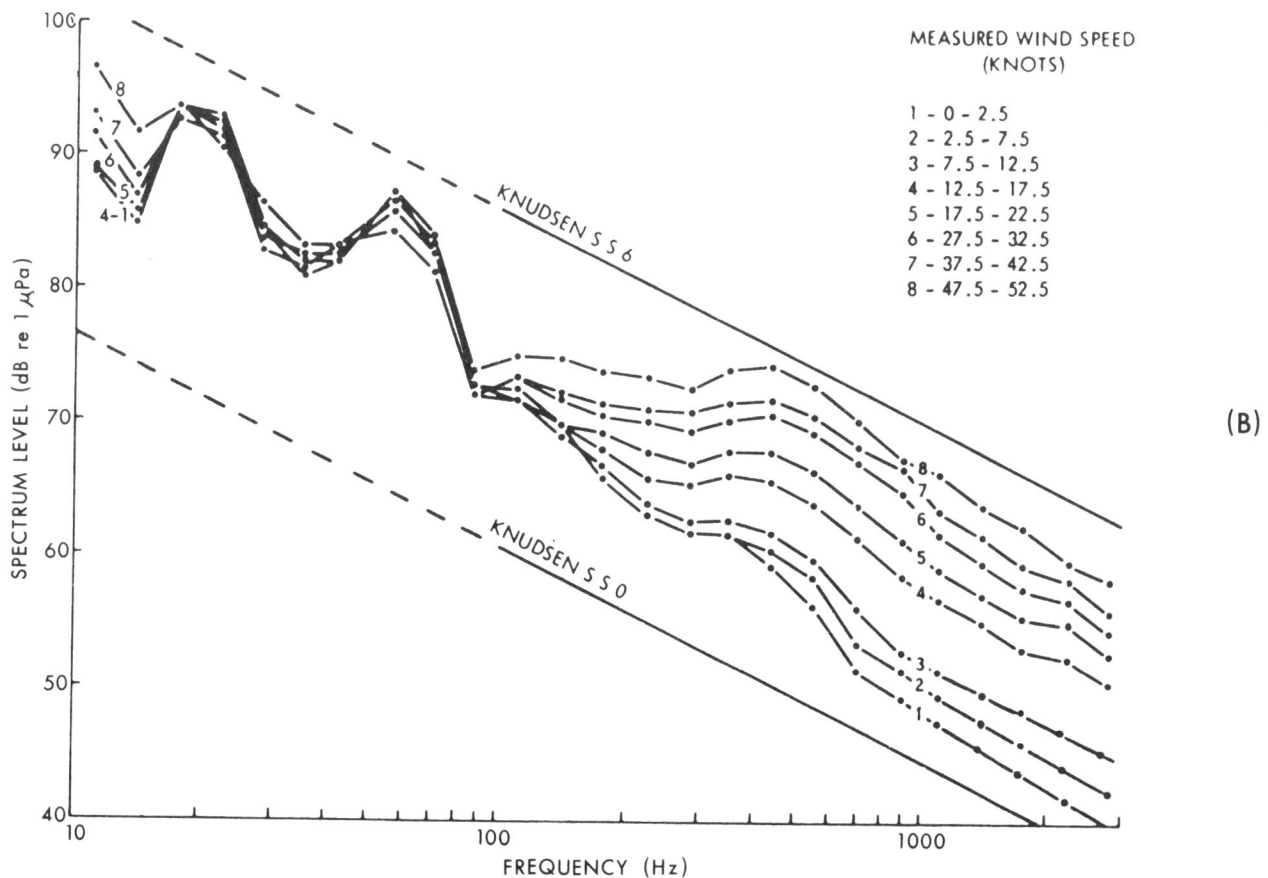

Figure 2-8. Effect of wind speed at different frequencies. (A) Variation of noise with time over a period of 27 days; the lowest cross-hatched curve is wind speed. (B) Noise spectra for various wind speeds. Reference 29.

Another area of high shipping density is the Mediterranean Sea. The measurements of Arase and Arase (34) at four locations in different basins of the Mediterranean were typical of those to be expected in an area of dense ship traffic at 50 Hz, but were about the same as those for the prevailing wind speed in deep open water at 800 Hz, where the wind, rather than shipping, is the principal noise source. The measured levels at the four locations are shown in Figure 2-9.

The importance of location is further illustrated by the observations at two deep Pacific Ocean sites by Wilson (35) at frequencies of 60 and 165 Hz. One site (Site A) was close to major trans-Pacific shipping lanes and storm tracks. The other site (Site B) was 450 miles south of A, and was therefore distant from shipping and storms. At 60 Hz at Site A, the noise was dominated by shipping all the time, with distant storms raising the 60 Hz level only from 83 to 89 dB. At 165 Hz, the stormy periods had a greater effect, ranging from 60 to 74 dB, but in the reverse direction; lower levels were observed during stormy periods as a result, it was claimed, of the lower speeds of ships during storms. At Site B, the 60 Hz noise levels were dominated by storm noise part of the time, ranging from 72 to 80 dB over a month's time, while at 165 Hz the levels were about 62 dB in calm periods and 66 to 74 dB during distant-storm periods.

In this frequency band, one should not rule out noise of man-made activity other than shipping. In the western Gulf of Mexico, where seismic exploration is actively carried on, the ambient background was found (36) to be dominated by the noise of seismic exploration, consisting of pulses that could be detected audibly virtually all the time at a deep-water location. No measurements appear to exist in the open literature on the noise background in and near a producing oil field where seismic exploration is no longer occurring.

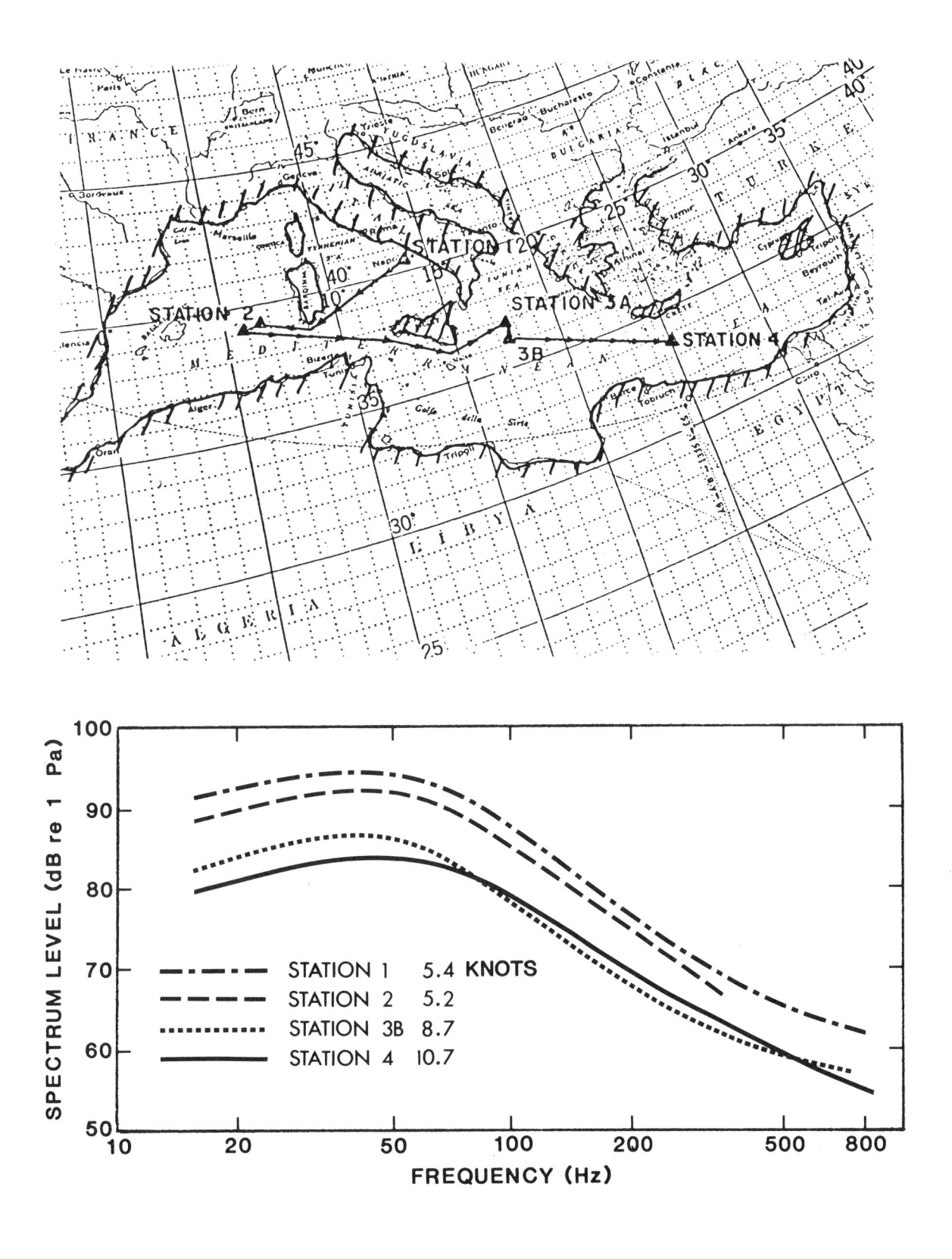

Figure 2-9. Noise levels at four stations in the Mediterranean Sea, averaged over three days at each station. Reference 34.

2-17

2.5 High Sonic Band (200 to 50,000 Hz)

This frequency band was explored during World War II by a group headed by the eminent acoustician, V. O. Knudsen. The results of measurements made in 14 bays, harbors and offshore areas off our east and west coasts, Hawaii and Great Britain were summarized in a series of curves (actually straight lines on a logarithmic frequency scale) now known as the "Knudsen Curves" that were first published in 1948 by Knudsen, Alford and Emling (37).

We should note at the outset that the locations at which these wartime measurements were made were in shallow coastal water, there being evidently little opportunity in wartime to make measurements in deep water far from shore. Yet the Knudsen curves have since been regularly used for making estimates of expected noise levels in deep water. However, at the low end below 1 kHz, they show no evidence of the flattening or a turnover of the spectra which subsequent deep water observations have revealed. Yet, the Knudsen levels--resulting from the first systematic investigation of ambient noise ever made--have proven over the years to be useful for noise level predictions at frequencies of 1 kHz and above.

Concerning these spectra it is said in this early report that:

> "the slope of the spectrum appears to be independent of wind and sea and averages -5 dB/octave. Experimental evidence indicates that random departures from this slope may occur, but usually the slope will not be more than -6 or less than -4 dB/octave. Neither noise level nor spectrum varies greatly with water depth so long as the water is sufficiently deep to prevent breaking of the waves. In deep water the average noise level and spectrum are essentially independent of the depth (20 to 300 ft) at which the noise is measured. There is, however, a difference in the character of the noise. Near the surface the noise from individual waves and whitecaps can be discerned, and the momentary variations in noise level are greater than at a greater depth."

Figure 2-10 shows the Knudsen curves as they were published in 1948, converted from a reference pressure of 0.0002 dyne/cm to 1µPa, with the curves dashed below 1 kHz to indicate caution in their use in this region, because of the possible presence of shipping noise or, in its absence, a change in slope or a reversal of slope of wind noise in this band as indicated by subsequent measurements.

The levels are, as the curves show, strongly dependent on wind speed, and have a slope slightly less than 20 dB per decade of frequency; in other words, the noise intensity falls off at slightly less than the inverse square of the frequency. Subsequent measurements tend to show a flat spectrum for wind noise between 200 and 800 Hz in both deep and shallow water. The data obtained by Perrone (29) show better correlation with wind speed than with sea state, perhaps because wind speed can be more definitely measured than can sea state.

A number of more recent observations may be cited. In 1954, using a surface-suspended hydrophone, Johnson (38) made noise measurements at five deep-water locations off the west coast of Mexico and Central America, and found levels confirming those of the Knudsen curves. Perrone (39) (29) made noise measurements for a one-month period with the Trident Vertical Array (see Figure 6-4), recording noise for 2 minutes every 2 hours, with the results already shown in Figure 2-8, and less dramatically in Figure 2-11. Particularly noteworthy is the dependence on the wind above 200 Hz and also at 11 Hz. Figure 2-12 shows a similar finding (40), but in the enclosed basin north of the island of St. Croix.

In the Southern Hemisphere, Burgess and Kewley (32) obtained measurements in deep water between Australia and New Zealand and found wind-dependent noise over the whole frequency range from 20 to 800 Hz with a flat spectrum between

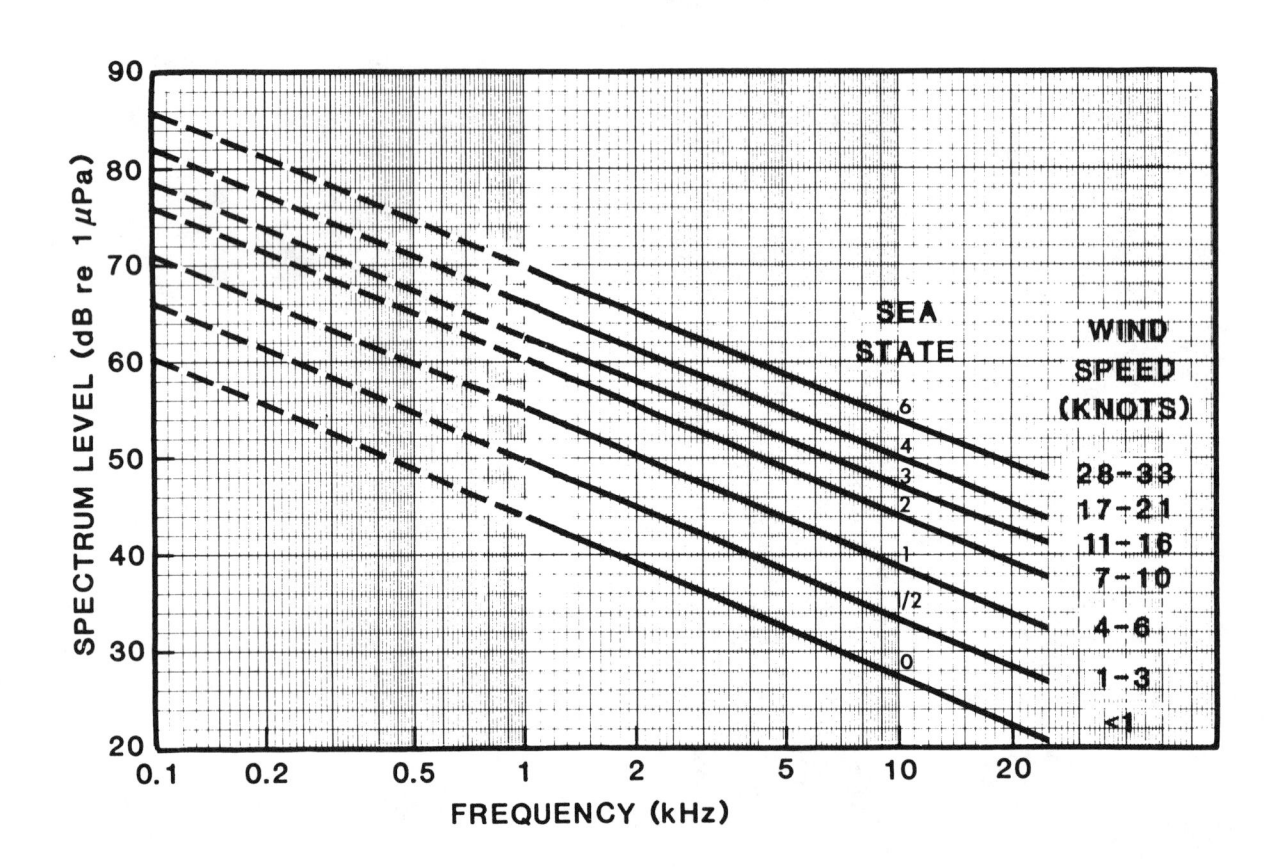

Figure 2-10. The "Knudsen Curves," published 1948. Reference 37.

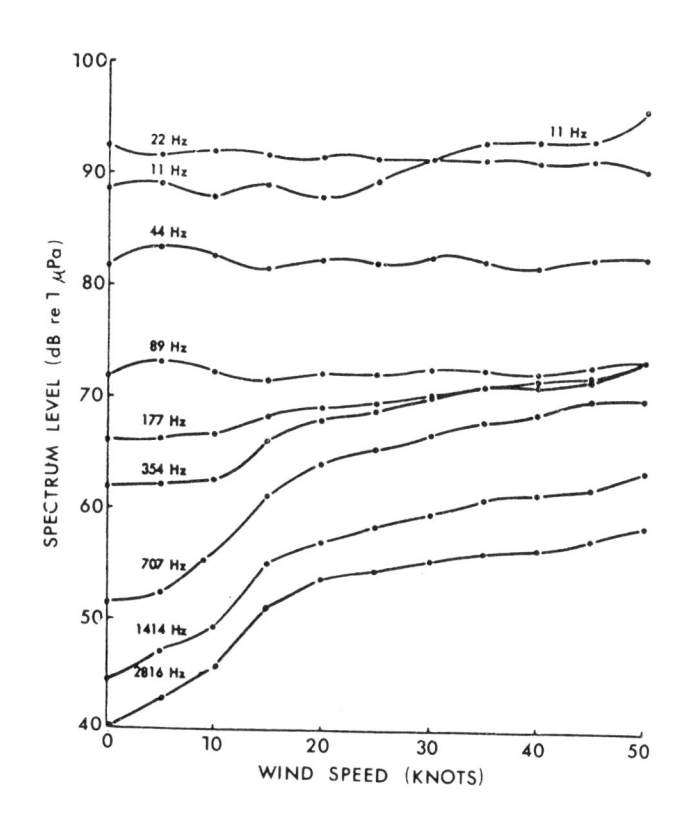

Figure 2-11. Noise spectrum level versus wind speed at different frequencies in the northwest Atlantic. There is a strong dependence on wind speed at high frequencies down to about 200 Hz and also at 11 Hz, with no wind dependence in between. Reference 29.

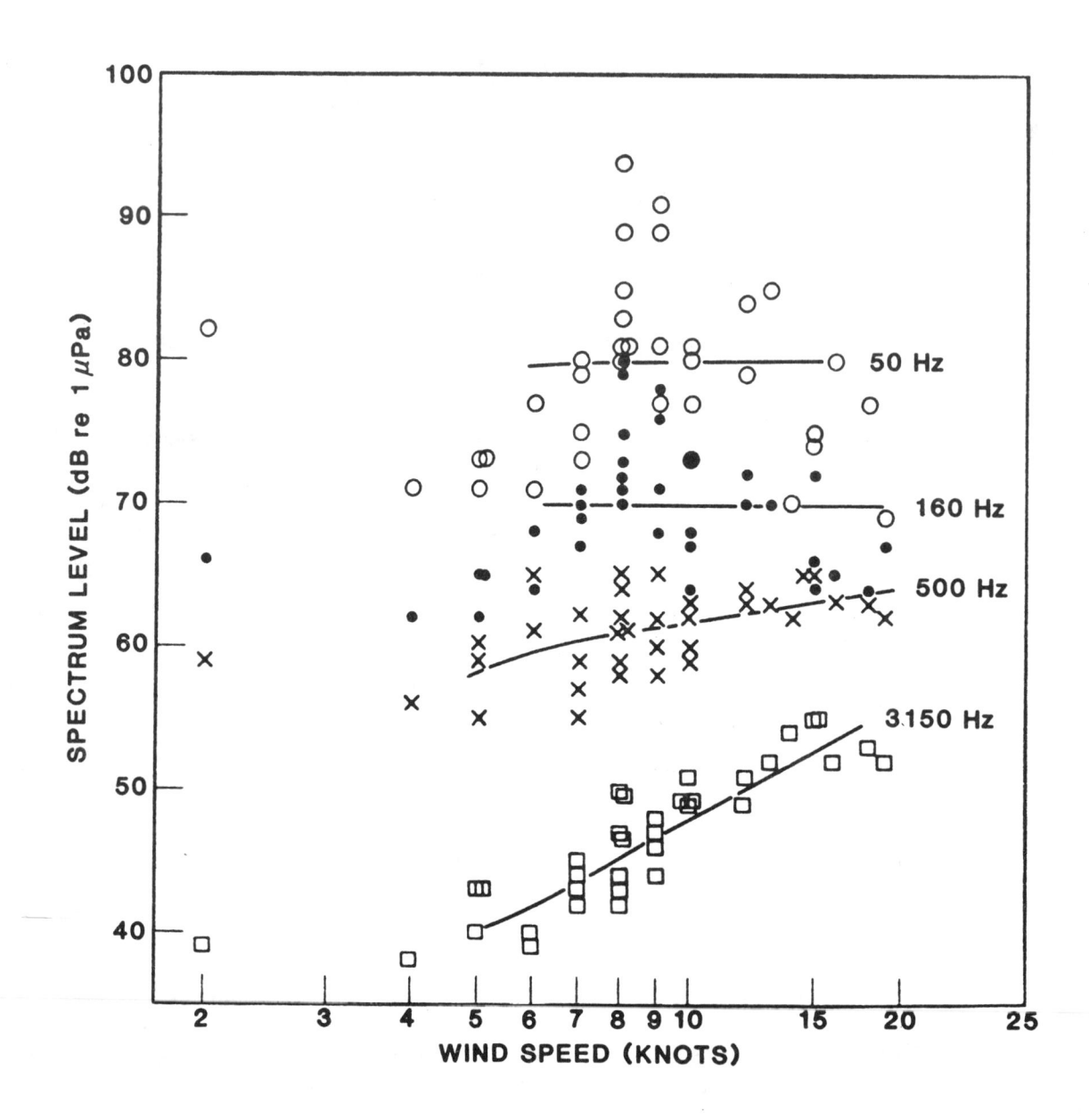

Figure 2-12. Variation of level with wind speed at various frequencies north of St. Croix. Reference 40.

200 and 800 Hz. Figure 2-13 shows the measured noise spectra from this paper, together with calculated spectra according to an empirical formula to be mentioned later on.

In a basin west of Guam -- the Parece Vela Basin -- wind noise was found to dominate shipping noise at frequencies as low as 150 Hz (41). This basin is large (1100 x 220 km) and deep at the measurement site (4800 m). Yet the absence of shipping, except at the edges of the basin, was believed responsible for the dominance of the wind down to such a low frequency. Figure 2-14 shows noise spectra as observed with a bottom-moored vertical array over a 5 to 10 day period. We note the sharp onset of ship noise toward lower frequencies below 150 Hz as well as a difference of 5 dB between a hydrophone at a depth of 3572 m and the other on the bottom at a depth of 4572 m. This depth dependence was attributed to near-shore ship traffic radiating into the deep sound channel by down-slope propagation (see Figure 2-20). We may also note the small effect of depth in the wind-dominated region of the spectrum.

A number of attempts have been made to fit an empirical expression to data in the wind-dependent part of the spectrum. The initial attempt was made by Crouch and Burt (42) to fit data previously obtained by Piggott and by Perrone for the northwest Atlantic. The selected expression was

$$NL = B(f) + 20 \, n \, \log_{10} V$$

where NL is the noise spectrum level in dB re 1μPa at frequency f, B(f) is the noise level at a wind speed of 1 knot, n is an empirical coefficient, and V is the wind speed in knots. If n = 1 the noise level increases as 20 log V, and the noise intensity will increase as the square of the wind speed. Table 2-1 is a listing of the coefficients representing measured data as reported for the northwestern Atlantic Ocean, off Guam, and two sets of data near Australia. The frequencies selected for this table are at octave intervals.

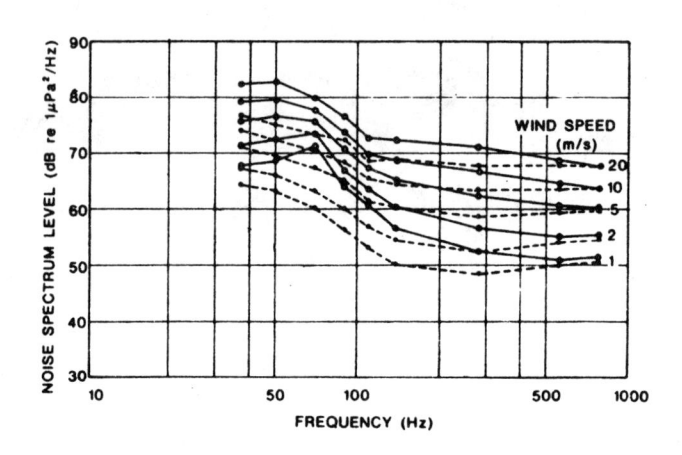

Figure 2-13. Noise spectra measured between Australia and New Zealand. Reference 32. The dotted curves are those calculated from an empirical expression with coefficients listed in Table 2-1.

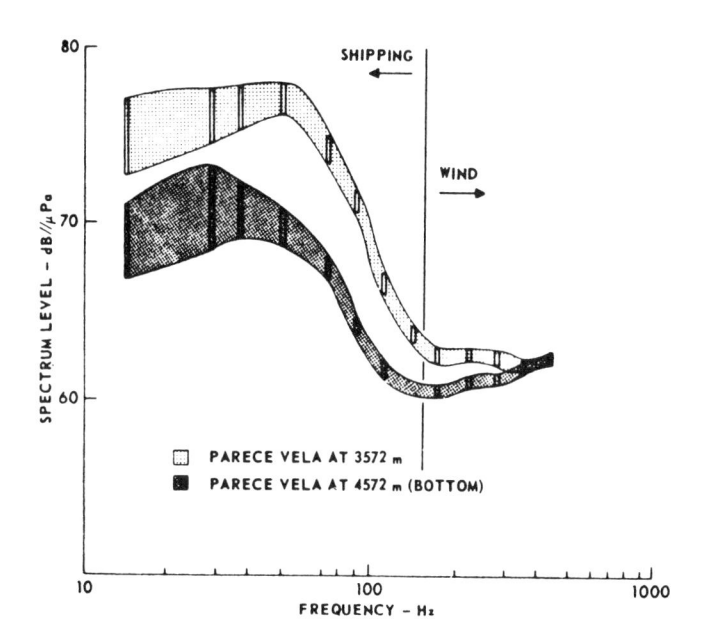

Figure 2-14. Spectrum levels at two depths in the Parece Vela Basin west of Guam. The measured levels at a bottomed hydrophone fell in the shaded band; those 1000 m above the bottom are in the unshaded band. The line at 150 Hz separates shipping from wind noise. Reference 41.

Table 2-1. Empirical Coefficients for Wind-Dependent Noise

Frequency (Hz)	(a) Crouch & Burt Ref. 42 (B)	(n)	(b) Shooter & Gentry Ref. 41 (B)	(n)	(c) Burgess & Kewley Ref. 32 (B)	(n)	(d) Cato Ref. 30 (B)	(n)
200	43	0.87	21	1.50	50	0.66	55	0.70
400	43	0.92	25	1.42	48	0.70	55	0.67
800	32	1.1	-	-	48	0.61	54	0.63
1600	30	0.96	-	-	-	-	52	0.60
3200	26	0.96	-	-	-	-	45	0.65

a. Crouch and Burt, analysis of Piggott's (67) and Perrone's (29) data off the Scotian Shelf and in the northwest Atlantic.

b. Shooter and Gentry, Parece-Vela Basin near Guam, hydrophone 1000 m above bottom.

c. Burgess and Kewley, from deep-water sites between Australia and New Zealand.

d. Cato, 40 sites off Australia.

No explanation is at hand to explain the discrepancies between the levels calculated by using the various coefficients. Even though it would be expected that wind noise would have the same level at the same wind speed in different areas as measured by different workers (in contrast with non-wind dependent noise), there are unexplained differences between the data given in the cited literature. Of the different values of n, a value of n = 1 (meaning that the noise intensity varies as the square of the wind speed) has an intuitive appeal, because the stress of the wind upon the sea surface, and the drag of most obstacles in a moving fluid, also vary as the square of the wind speed.

If the noise level is indeed related to wind speed, as is certain to be the case at kilohertz frequencies, then it becomes possible to use a hydrophone as an anemometer for wind speed measurements at remote underwater locations. Thus, an instrument called WOTAN (Weather Observations Through Ambient Noise) is basically a hydrophone which measures noise at 4.3, 8 and 14.5 kHz (43). The instrument distinguishes rain noise from wind noise; rain noise, being "white", has the same level at all three frequencies while wind noise does not. Extensive trials of this underwater anemometer were conducted in the mid-Atlantic Ocean by Shaw and others (44). At 5 kHz it was found that the wind speed was related to the measured noise level through the relation $20 \log V = 1.01 (NL) - 30.4$, where V is the wind speed in knots and (NL) is the noise spectrum level in dB re $1\mu Pa$, with an accuracy of ± 5 knots.

2.6 Ultrasonic Band (> 50 kHz)

At frequencies from 50 to 200 kHz, depending on wind speed, thermal noise begins to dominate the noise background. Thermal noise is the noise of molecular bombardment. It is the analog of the Nyquist electrical noise in the radiation resistance of a radio antenna and places a limitation on hydrophone sensitivity at high frequencies. This form of noise was first investigated theoretically by Mellen (45), who found through considerations of classical statistical mechanics the expression

$$NL = -15 + 20 \log f$$

for the spectrum level NL in dB re 1µPa of the thermal noise at frequency f in kHz. This expression is plotted in Figure 2-15 along with extrapolated Knudsen curves. Some 30 years later, Sullivan and Kemp (46) reviewed Mellen's work and made some minor corrections to the derivation. Experimental verification, within 3.6 percent, of the above expression was obtained by Ezrow (47) using an ingenious experimental method in the laboratory.

No measurement work in the real ocean has been done in this frequency range, except for the measurements of Anderson and Gruber (48) at 30, 90 and 150 kHz in the ports of San Diego, Long Beach in California, Balboa and Christobal in the Pacific Canal Zone, and Norfolk, Virginia. These locations were found to be extremely noisy, and showed great variability from port to port. The average levels in these ports was some 20 dB higher than the Knudsen extrapolated levels for sea state 6. Surprisingly small differences were found between day and night; the lower levels due to industrial activity during the night were evidently compensated by higher noise due to snapping shrimp. Comparing the various ports, there was a general tendency for the noise levels to increase with decreasing latitude, as would be expected from a greater abundance of shrimp in lower latitudes. From this data we conclude that the thermal noise limit at 30, 90 and 150 kHz is far lower than the noise levels prevailing in busy harbors.

No measurements in this ultrasonic band in deep, quiet open water appear to have been made.

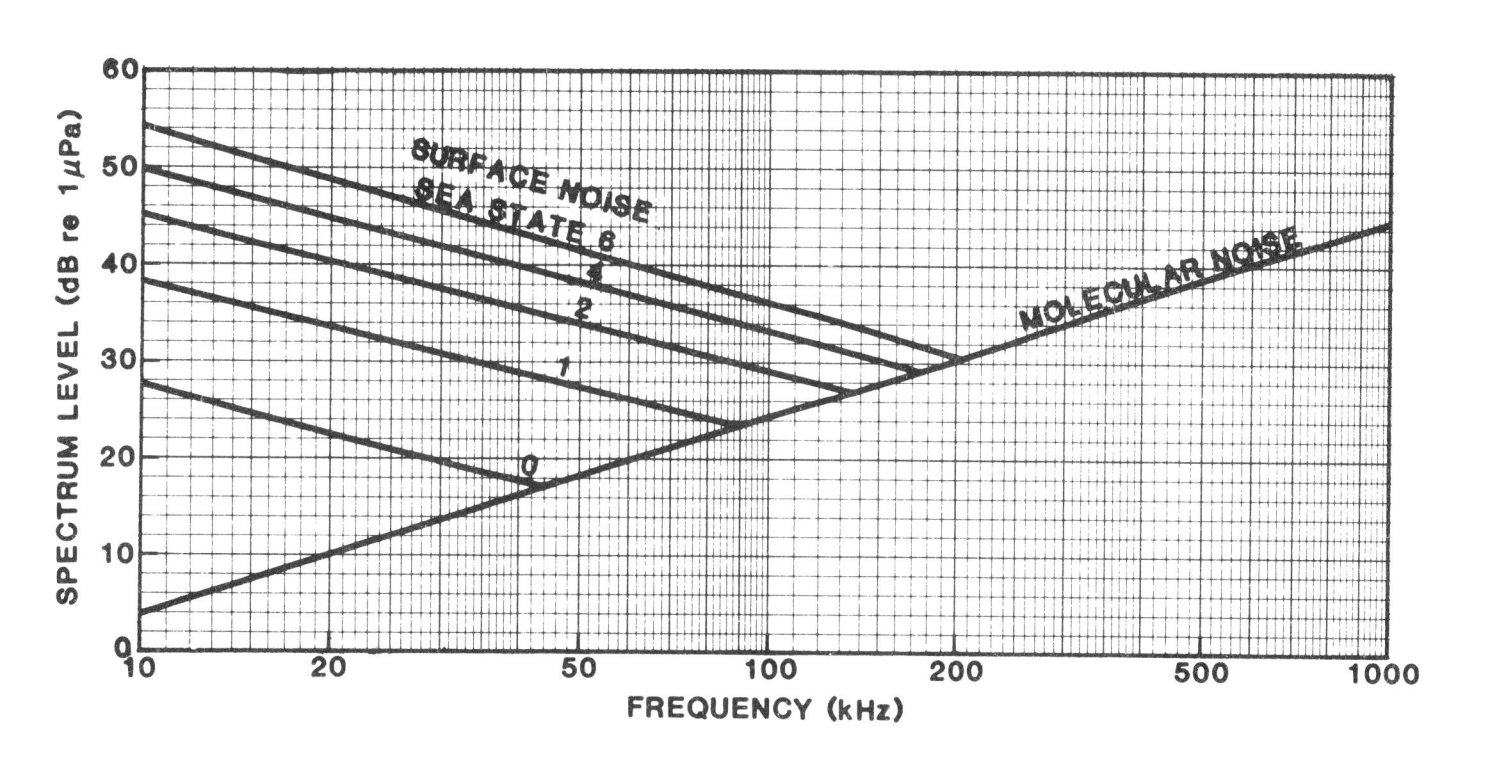

Figure 2-15. Surface noise and molecular noise in the band from 10 kHz to 1000 kHz. Reference 45.

2.7 Processes of Surface Noise Generation

The processes by which the wind causes the ambient noise in the sea have been much speculated upon by theoreticians. Doubtless, different processes are dominant in different portions of the overall frequency band from 1 to 50,000 Hz. These processes are here discussed together for convenience. A general theory of surface noise, covering all mechanisms and containing a large number of integral expressions, has been formulated by Li (49).

A number of source mechanisms may be identified: (1) wind turbulence, (2) surface motion, (3) wave interactions, and (4) spray and cavitation.

1. <u>Wind Turbulence</u>. Because the wind is turbulent and sea is rough, the turbulent pressures of the wind blowing across the rough sea surface appear as noise pressures in the sea below. The mechanism is the same as that which produces the rough sea surface itself. In a Russian paper (50), the spectrum of the noise is expressed in terms of the surface wave spectrum and a favorable comparison is made between theory and at-sea measurements. The theory is summarized and extended in a paper by Wilson (51) and correspondence between the theory and certain observations is obtained in the infrasonic spectral region of the spectrum. Figure 2-16 shows theoretically computed spectra from this paper for wind turbulence and wave interaction described below (No. 3).

2. <u>Surface Motion</u>. The motion of the surface itself can cause noise originating at the high frequency capillary patches of turbulence distributed randomly over the ocean surface. Kuo (52) investigated theoretically the pressures produced by surface motion, while Yen and Perrone (53) derived the theoretical spectrum of noise at a depth in the sea from the spectrum of the waves at the surface.

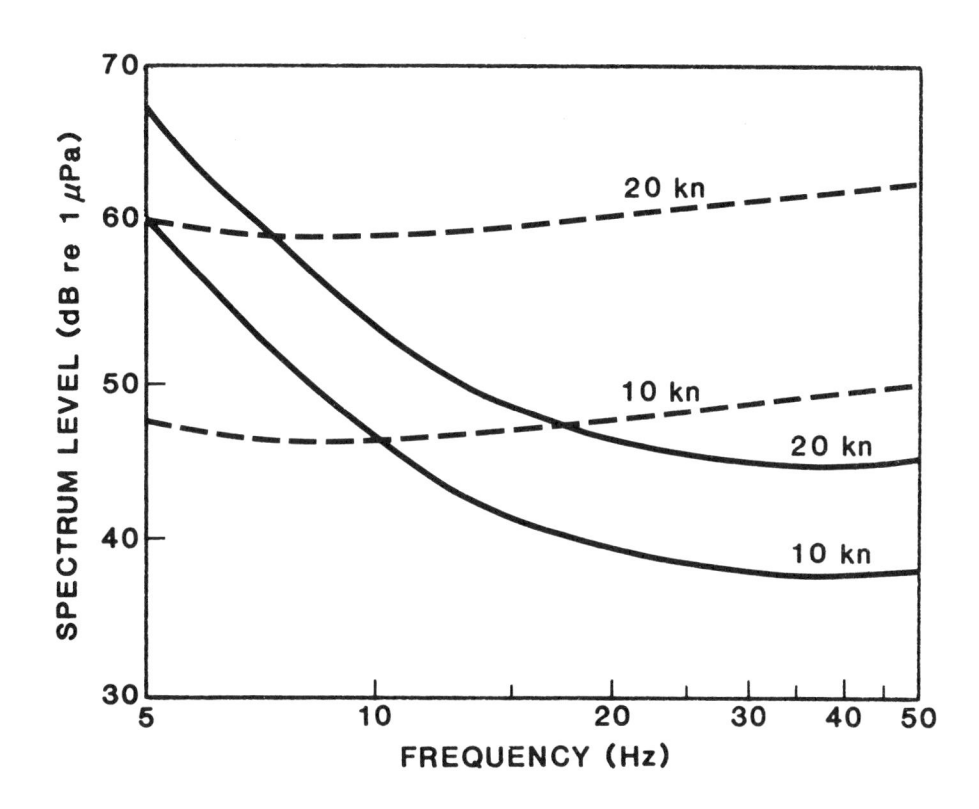

Figure 2-16. Theoretically computed spectra for the two principal sources of wind-induced low-frequency noise: wave interactions (solid curves), and atmospheric (wind) pressure fluctuations (dashed curves). Reference 51.

3. **Wave Interactions.** When two surface waves of the same wavelength travel in opposite directions, a standing wave is produced. In such a case, the pressure does not decrease with depth, as would occur with a single wave, but remains constant with depth. Longuet-Higgins (54), in a classic paper, showed that in water of constant depth, as a result of second order terms in the hydrodynamic equation, two oppositely traveling waves of equal amplitude interact to produce a pressure given by

$$p = \frac{-\rho a^2 \omega^2}{2} \cos 2\omega t$$

where p is the pressure relative to the mean pressure at some depth, ρ is the static fluid density, a is the amplitude of the two waves, and ω is 2π times the frequency of the surface waves. We note from this expression that the pressure p is independent of depth, has an amplitude proportional to the square of the wave amplitude, and has a frequency twice that of the surface waves. Marsh (55) in 1963 first applied this result to ambient noise in the sea and derived the expressions

$$p_f^2 = 94 \, H^{6/5} f^{-3}, \quad 1 \leq f \leq 13.5 \text{ Hz}$$
$$= 2.9 \, H^{6/5} f^{-5/3}, \quad f \geq 13.5 \text{ Hz}$$

where p_f is the ambient noise pressure in a 1 Hz band in dynes/cm^2, H is the wave height (crest-trough) in feet, and f is the frequency in Hz. Good agreement was claimed with the Knudsen spectra (Figure 2-10). Various subsequent theoreticians have extended the Longuet-Higgins-Marsh theory, including Brekhovskikh (56), who investigated noise generation by a continuous spectrum of waves, Harper and Simpkins (57), who extended the theory to large amplitudes, Hughes (58), who used a more modern surface wave model and obtained agreement with measured data below 10 Hz, and, more recently, Lloyd (59), who used an entirely different theoretical approach to the problem. The infrasonic noise produced by non-linear interaction of surface waves is said by Brekhovskikh (60) to be radiated up into the atmosphere as well as down into the sea; the noise of intense storms, propagating through the atmosphere, is said by Brekhovskikh to be observable on dry land at distances of up to a thousand miles.

4. **Spray and Cavitation**. Cavitating air-vapor bubbles have been hypothesized by Furduev (61) as a source of noise. In this theory the sea near the surface contains air bubbles which are assumed to grow by a process termed "rectified diffusion" under the changing pressures of the waves and then to collapse, producing sound in the same way as the bubbles about a cavitating propeller or sound projector. In support of this hypothesis is the fact that ambient noise in the absence of distant shipping noise has a peak or a plateau in its spectrum at about the same frequency as the noise of a cavitating propeller. In a paper by Wilson (62) the spray of bubbles from whitecaps was advanced as a source of noise, using the theory of Franz (63) for the noise generation by the impact of droplets, finding agreement with data reported by Morris (64). A thorough recent study of the underwater sound made by breaking waves has been made by Kerman (65). However, this source of noise does not account for the wind-dependent noise at low wind speeds where whitecaps are absent (below about 10 knots).

2.8 Shallow Water

In shallow water, in the absence of local shipping and biological noise, wind noise dominates the noise of distant shipping over the entire frequency range. The reason for this is that the deep favorable propagation paths traveled by distant shipping noise in deep water are absent in shallow water; in other words, the poor transmission in shallow water screens out the noise of distant ships and allows locally generated wind noise to dominate the spectrum at all frequencies.

Two sets of observations may be cited confirming the dominance of the wind in shallow water. Figure 2-17A shows Russian data for a bottom-anchored hydrophone at a depth of 200 fathoms at an unstated location (66). The dashed lines represent the Knudsen curves. Figure 2-17B shows the average noise spectra reported by Piggott (67) for a one-year period of measurement with two bottomed hydrophones in 20 and 28 fathoms of water on the Scotian Shelf. Both sets of spectra are seen to be dependent on wind speed at all frequencies.

Figure 2-18 shows spectra obtained at six locations, each for a 24-hour period off the coast of New Guinea at a depth of 18 m by Wylie (68). Included are spectra of biologic noise (Section 7.2) and rain noise (Section 7.4.1) as observed in this area.

However, in an area of extremely dense ship traffic, the noise of ships at short and moderate ranges may still dominate the low frequency spectrum, even though the water is shallow. This was the case in the North Sea, where Browning and others (69) found levels at frequencies from 30 to 200 Hz that were appreciably higher than those of Piggott (Figure 2-17B) on the Scotian Shelf. These high levels, which were independent of wind speed, can be attributed to the heavy ship traffic in the North Sea plus the noise of drilling rigs in the area. A more extensive series of noise measurements, along with accurate wind and shipping data at a research platform in the North Sea and on a ship in the Baltic, has been published by the German investigators Wille and Geyer (221).

An interesting comparison between deep and shallow water noise was made by Arase and Arase (70), who compared the noise received by hydrophones at depths of 30 and 900 fathoms located a distance of 3.3 miles apart on the

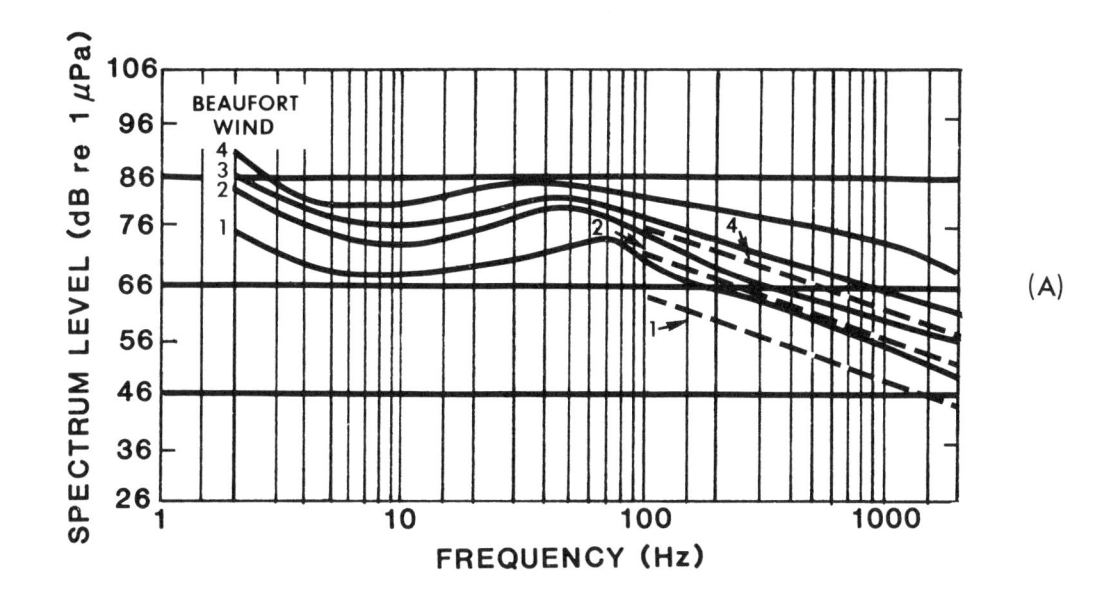

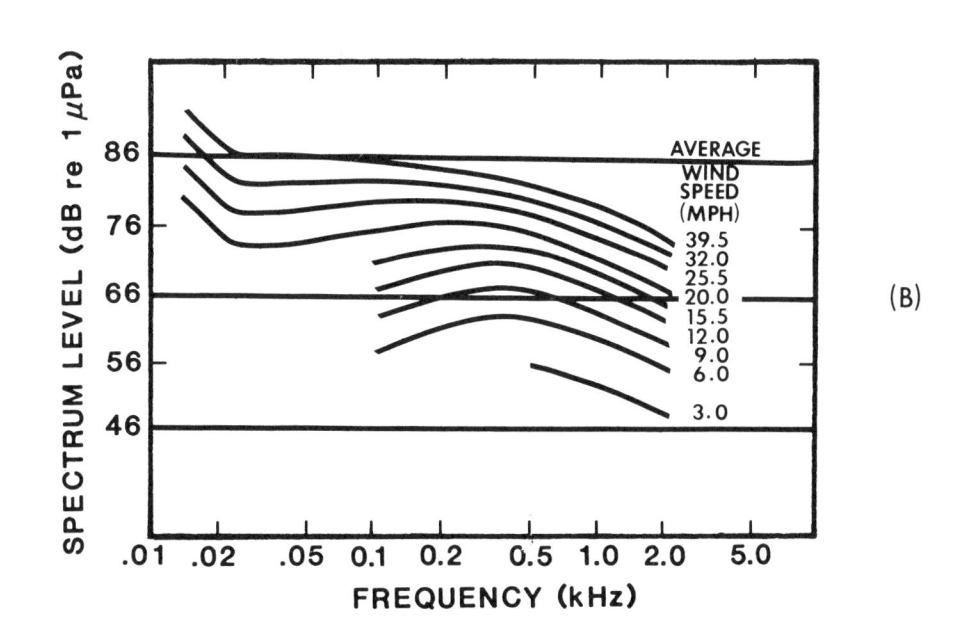

Figure 2-17. Shallow water ambient noise spectra, with Beaufort wind number as a parameter. (A) Russian data in 200 fms of water at an unstated location, Reference 66. (B) Scotian Shelf, 28 fms, Reference 67.

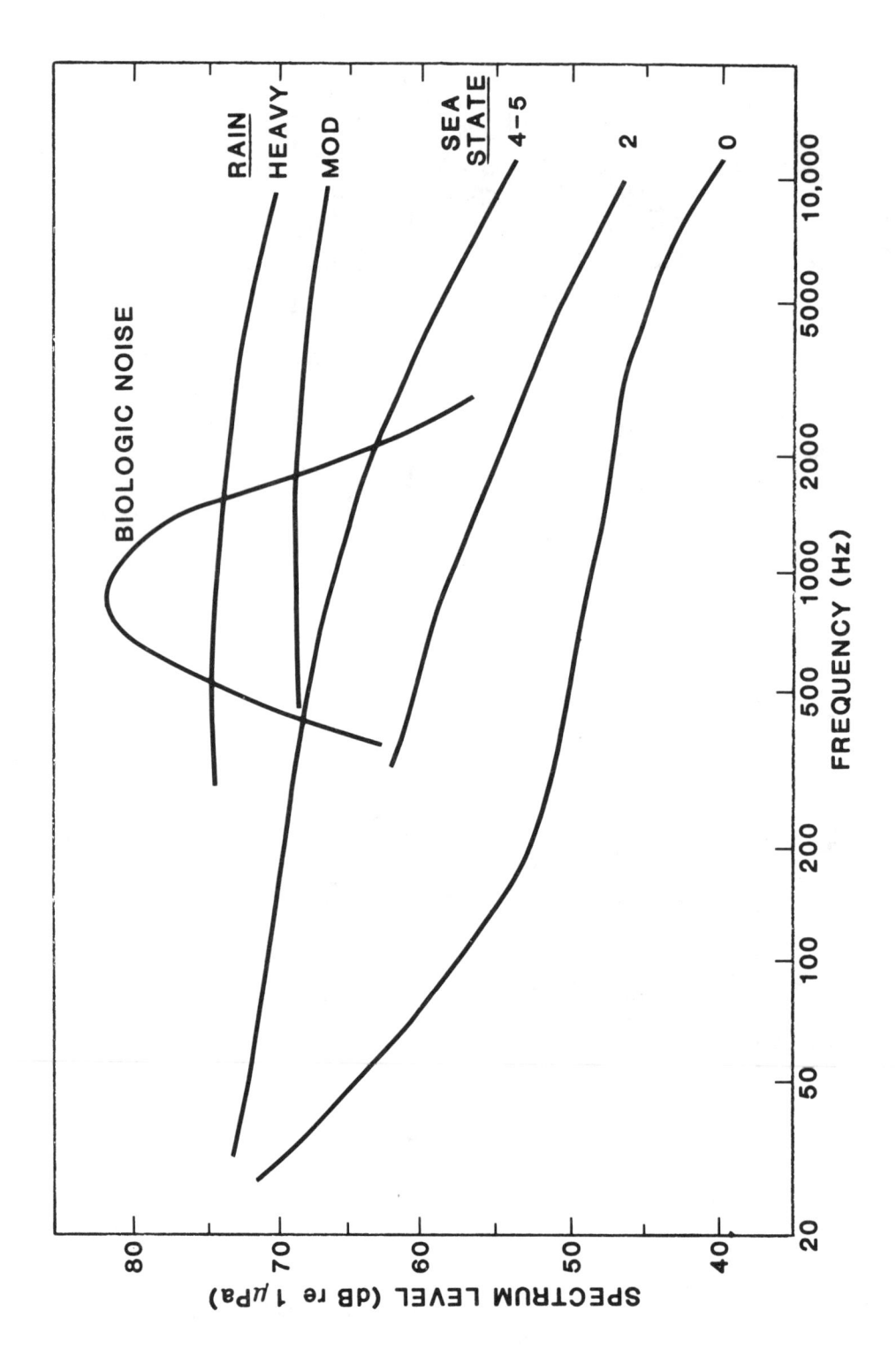

Figure 2-18. Noise spectra at six locations in the Coral, Solomon and Bismark Seas. Reference 68.

slope south of Bermuda. Propagation measurements showed that the level of a signal from a 100-Hz source at ranges from 30 to 300 miles was 14 dB less at the shallow hydrophone than at the deeper one. As a likely result of this propagation difference, the noise levels over the frequency range 22 to 715 Hz were higher at the deeper hydrophone, with a maximum different occurring at 50 Hz. Figure 2-19 shows that this difference diminishes with increasing wind speed, as the wind contributes more and more to the noise at both hydrophones as the wind increases. The correlation of level with wind speed was high ($\rho > 0.5$) over the entire range of frequency (22 to 715 Hz) at the shallow hydrophone, but was high only at 500 and 715 Hz at the deep unit. An increasing correlation of level with wind speed with increasing frequency was also reported in a Russian paper (71). All of this is consistent with the view that only wind noise occurred at the shallow unit, whereas the deep hydrophone picked up both wind and distant shipping noise.

Even in the confined waters of Narragansett Bay, 1/3-octave band levels at 530 Hz were strongly correlated with both wave height and wind speed (72). At this location a fully developed sea and an equilibrium sea condition was found to exist (i.e., the rate at which energy is transmitted to the waves is equal to the rate at which it is dissipated) when the wind blew steadily for 1/2 hour. With an increasing or decreasing wind, the wind speed was a more significant variable than wave height for estimating the noise level.

2.9 Effects of Propagation on Noise

The propagation of sound in the sea affects the ambient noise background in a variety of ways. The greatest effects occur for distant shipping noise because such noises originate at long distances from the receiving hydrophone. One such effect has been called the "down-slope conversion effect" where sound originating the shallow water of the continental shelf is introduced into the Deep Sound Channel (DSC), wherein it travels to great distances with

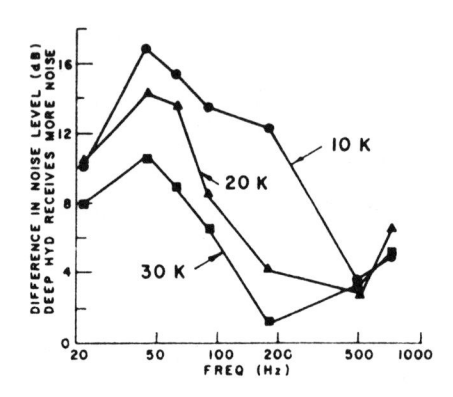

Figure 2-19. Difference in noise level between a shallow (30 fms) and a deep hydrophone (900 fms) 3.3 miles away off Bermuda. The deep hydrophone receives more noise by the amounts given in dB. Reference 70.

relatively little loss (73). The effect has also been called the "megaphone effect." The "conversion" is illustrated by Figure 2-20. By this process, a deep receiver receives more sound from distant coastal ship traffic than it would in water of uniform depth. At the same time, the directional pattern is affected. Measured noise patterns (Section 5.4) show little or no evidence of the dip or notch in the pattern in the horizontal direction that would be expected in deep water of constant depth. The notch in the pattern is believed to be filled in along the horizontal by noise traveling down the DSC from coastal sources.

A similar effect can be produced by a changing velocity profile. Thus, the transition in the velocity profile across the Gulf Stream from warm water on the west to cold water on the east causes more sound to be received at deep hydrophones at Bermuda than would occur with a uniform velocity profile, and, in addition, changes the directional character of the noise. At a location midway between Cape Hatteras and Bermuda, the vertical beamwidth of noise was found by Wales and Diachock (74) to be $\pm 20°$ in the range 45 to 100 Hz, with peaks occurring at 0° and $\pm 15°$. The peak at 0° was attributed to open-ocean shipping in the deep water southeast of the Gulf Stream.

Another example of down-slope conversion may be cited (75). Explosive shots at a depth of 1 to 2 m, fired off the California coast, were received at deep hydrophones at Eniwetok, Midway and Oahu in the Pacific. The strongest signals were those fired at locations near the edge of the continental slope in water depths of 90 to 275 m. Those fired in more shallow water underwent a greater number of bottom reflections, causing them to be weaker, while those fired in deeper water were weaker because they could not be "converted" into the DSC.

In the shallow water of the Gulf of Maine (20 fms), the measured noise levels were found (76) to be unusually low and to depend on wind speed at all

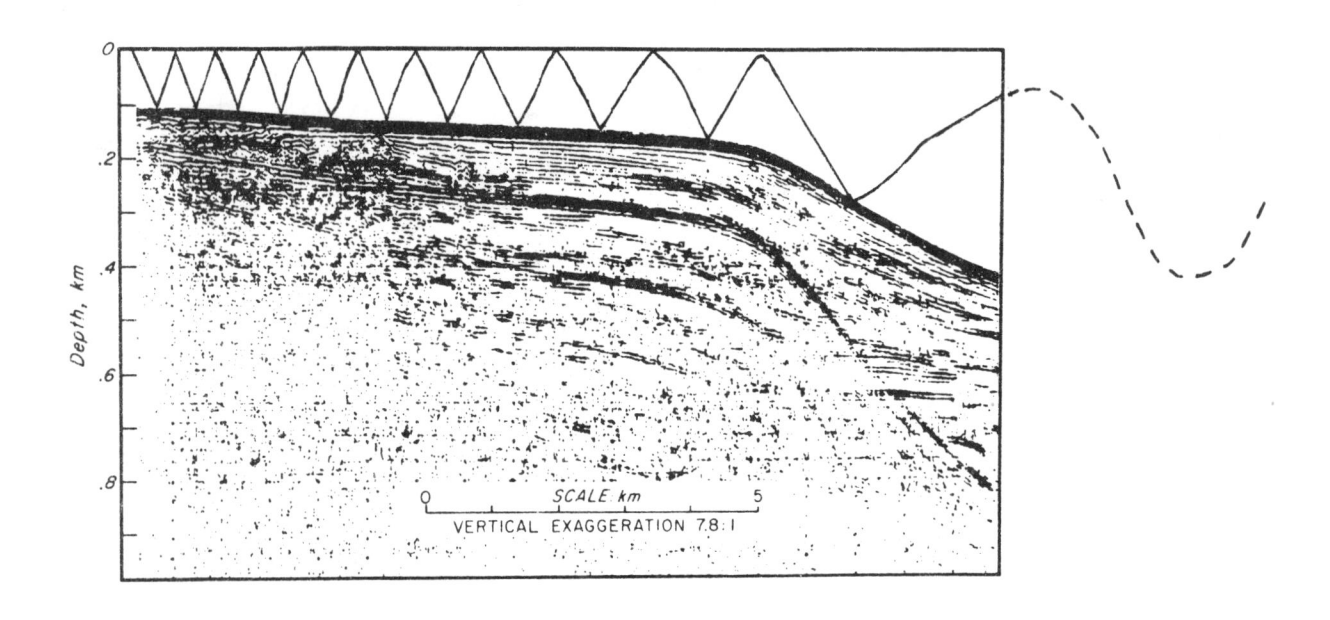

Figure 2-20. Cross-section of the continental shelf and slope off Pt. Arena, California, showing how a ray leaving the source at an angle of 15° becomes a ray in the deep sound channel. Reference 75.

frequencies. This was attributed to a sparsity of shipping and poor transmission in an area having a hard rock bottom. In this area a vertical-component velocity pickup, or "geophone," indicated ground velocities, when converted to pressure in the water, that were of the same order of magnitude as the noise pressures in the water; unfortunately, no horizontal-component data were collected. On the other hand, at another site over a sand bottom where the transmission was better, higher levels and a strong wind speed dependence was observed; in this case distant shipping at ranges of many tens of miles contributed to the noise. Thus, it is clear that at low frequencies in shallow water, a correlation must exist between the transmission and the level of the noise background. Poor transmission serves as a blanket or shroud against the noise from far-distant sources.

There can be no doubt that the bottom, instead of the water, in shallow water is the major sound conducting medium, especially at frequencies below the cut-off frequency for transmission in the water. A three-component geophone placed on the bottom (77) near the entrance of Narragansett Bay for noise measurements showed that the dominant factor influencing the noise levels was the tidal current; more noise occurred during periods of peak tidal current, both on an exposed and a buried geophone. Doubtless, current turbulences were the cause, but the mechanism of noise generation in this case, whether distant or local, is not clear.

The local, as well as the distant, bathymetry may affect the noise. For example, if a bottomed hydrophone lies in a depression or a valley in the bottom topography, it will, in all likelihood, be screened off to some extent from distant sources of noise and be quieter. Contrariwise, if it is placed on a peak or ridge, or on a seamount, the location is likely to be more noisy by being elevated up into the DSC. When on the bottom, the mere proximity to the lossy boundary will doubtless affect the noise.

Other examples of the effects of propagation on noise will be noted in Sections 4.5 and 5.4.

2.10 Summary

In summary, it may be said that the wind, by whatever process, creates the ambient background of the sea over the entire useful frequency range whenever the noise pollution produced by ships is absent. The spectra of wind noise at high and low frequencies is, however, different, and there is a region of lesser slope between them. This suggests the prevalence of two different processes of noise generation at low and high frequencies, as suggested by Figure 2-21. Which of the various processes these two are, or whether some entirely different noise mechanism is responsible, has yet to be determined. In short, the present status of these various mechanisms is in doubt, and the range of validity, if any, of the various processes is unknown. As suggested in a review paper by Li (49), a detailed evaluation of existing wind noise data, plus additional field work and studies on air-sea interactions, ocean dynamics and turbulence processes is needed.

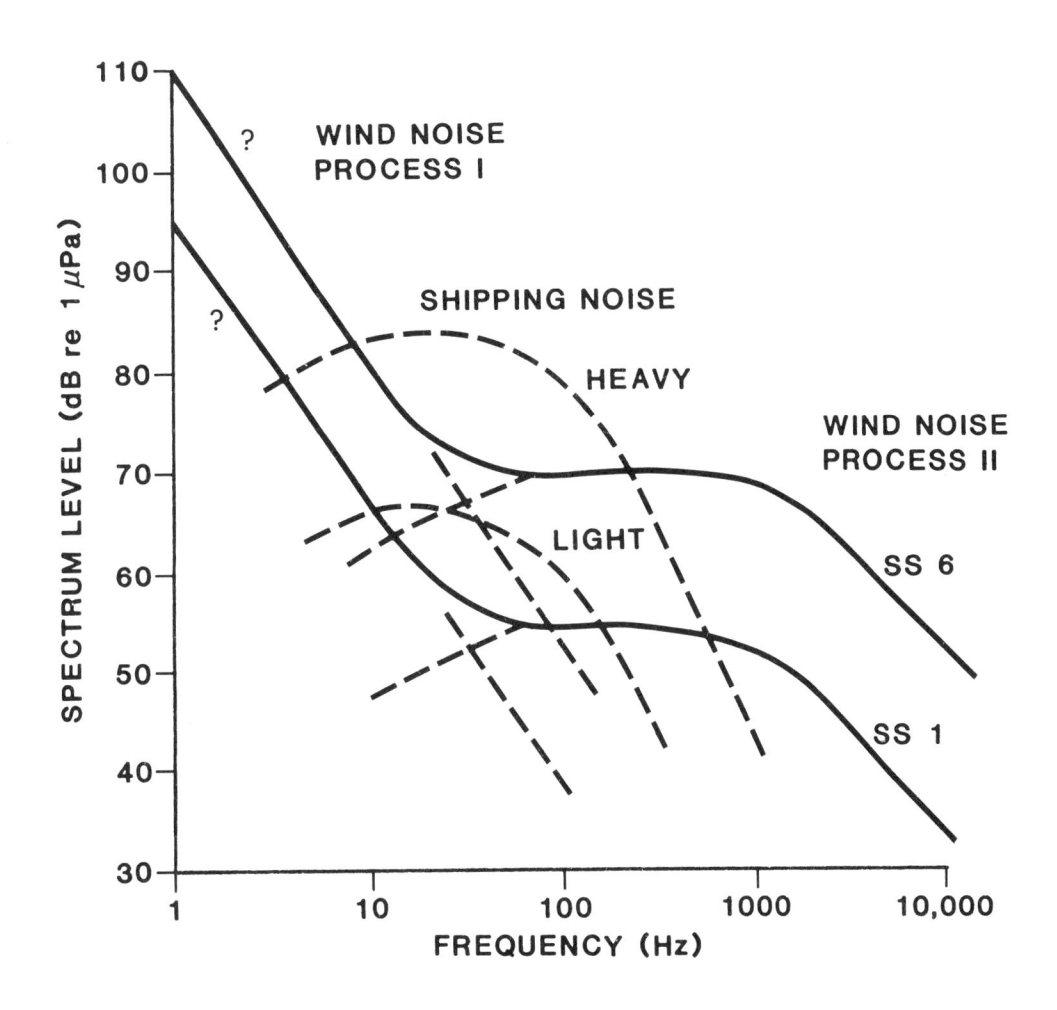

Figure 2-21. Conceptual view of noise spectra over the range 1 to 10,000 Hz, showing shipping noise plus two noise processes at low and high frequencies.

3. VARIABILITY OF AMBIENT NOISE

3.1 Introduction

Like everything else in underwater sound, the level of the ambient noise background at a fixed location varies with time. This time variability covers a wide scale, from the very fast, such as the transients of breaking waves, to the very slow, such as long-term changes of ship traffic or long-term changes in weather and climate. Such changes reflect the variability of the sources of noise, as well as of the sea as a medium of sound propagation.

The variability of noise can be expressed in terms of its <u>fluctuation spectrum</u>, giving the fluctuation power per unit frequency band, as a function of frequency. Such fluctuation spectra are useful for revealing the sources of noise fluctuations and for providing basic data for theoretical investigations.

However, in the literature, some other, and often more meaningful fluctuation parameter is reported, such as the standard deviation of a succession of noise samples. Yet the existence of long-term variations--seasonal or longer--means that all of our noise statistics are, strictly speaking, non-stationary over long periods of time.

In considering variability there are several time intervals involved (see Figure 3-1). One is the averaging time t_1 of a single sample, as would be read at the output of a spectrum analyzer or as the mean of a number of short-term digital noise samples. The second time interval, t_2, is the interval between the sample averages of duration t_1; for contiguous or consecutive samples, $t_2 = t_1$. The third interval, t_3, is the overall data recording period, or ensemble duration. As an example of these time intervals, suppose we were to take one-minute samples of noise at hourly intervals over a period

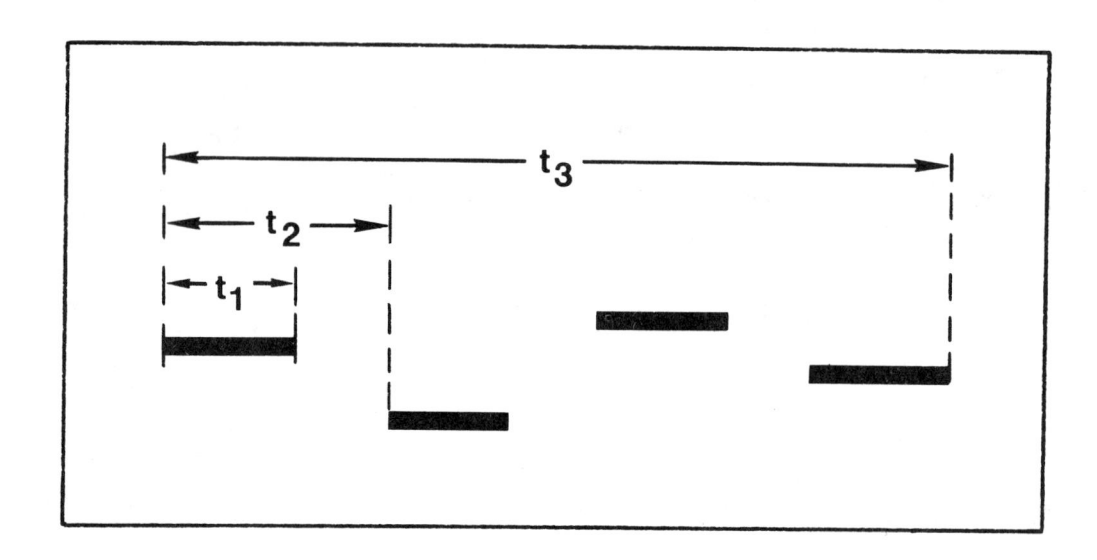

Figure 3-1. Time scales for variability of four noise samples. t_1, sample duration; t_2, interval between samples; t_3, ensemble duration. The four time intervals are necessary for interpretation of the derived statistics of a long series of such samples.

of one week; then t_1 would be 60 seconds; t_2 would be 1 hour, and t_3 would be 7 days or 168 hours; the total number of samples analyzed would be then t_3/t_2 = 168. Although these considerations seem elementary, they are of direct interest to the user of variability data, but unfortunately are seldom all stated, or else are deeply buried, in published reports.

3.2 Noise Variability in Deep Water

In the following, we will divide the variability of noise into three time scales: (1) <u>short period variations</u>, covering time periods from a fraction of a second to a few minutes, (2) <u>medium period variations</u> extending from periods of a few minutes to a few weeks, and (3) <u>long period variations</u>, covering seasonal and longer-period changes.

3.2.1 Short Period Variations

By means of probability density analyses of data in one deep-water and two shallow-water areas, the ambient background was long ago found by Calderon (78) to have a Gaussian amplitude distribution over short periods of time. This finding has been repeatedly confirmed by a number of subsequent investigators when short samples of noise have been examined. For example, some 2000 to 3000 samples of noise taken at 30-millisecond intervals were found by Arase and Arase (79) to have Gaussian amplitude statistics. Also, the noise at the output of a towed line array after having been passed through a processor of variable bandwidth and integration time, was found by Urick (80) to have a Chi-square distribution, in agreement with the theory for Gaussian noise at the input of such a processor.

3.2.2 Medium Period Variations

The statistics of noise samples averaged over longer time periods have been commonly reported in terms of the standard deviation, σ, of the measured levels relative to the mean for some averaging time (t_1) over a total observation period (t_3). For example, at a location northeast of New Zealand in the South Fiji Basin, 1/3-octave levels of one-minute samples were found by Bannister and others (81) to have a σ between 5.5 and 7 dB at 10 Hz, decreasing with frequency to less than 2 dB above 200 Hz, as a result of the fact that there was little or no variation of wind speed during the observation period. In this work a gradual decrease of σ with frequency was observed between 10 and 100 Hz, presumably as the number of line components increased in the 1/3-octave analysis bandwidth and as the relative importance of wind noise increased.

For another area, Figure 3-2 shows cumulative distribution curves at five depths of 1/3-octave 1-1/2 minute noise samples recorded by Urick, Lund and Tulko (40) at hourly intervals over a 2-1/2 day period at a location north of St. Croix, Virgin Islands--an area in which passing commercial ship traffic at distances of the order of 20 miles is common. In this study, t_1 was 1-1/2 minutes, t_2 was 1 hour and t_3 was 2-1/2 days. At 50 Hz, for the 114 data samples, σ was of the order of 7 dB, while at 3150 Hz, σ was equal to about 6 dB, corresponding to the changes in wind speed during the observing period. During a subsequent overnight period at the same location (82), based on 18 samples, σ was similarly equal to 6 to 8 dB at 50 and 125 Hz where shipping noise predominated, but was only 2 to 3 dB at 5000 Hz because of the more constant wind speed over the shorter data-taking period. No depth dependence was found at 2000-foot intervals down to 12,000 feet in 13,600 feet of water, in agreement with the above-cited work (81) in the South Pacific at four depths down to 2300 m in 4000 m of water.

Variability occurs in narrow bands as well as in 1/3-octave or broader bandwidths. For example, Figure 3-3 shows the narrowband variability at

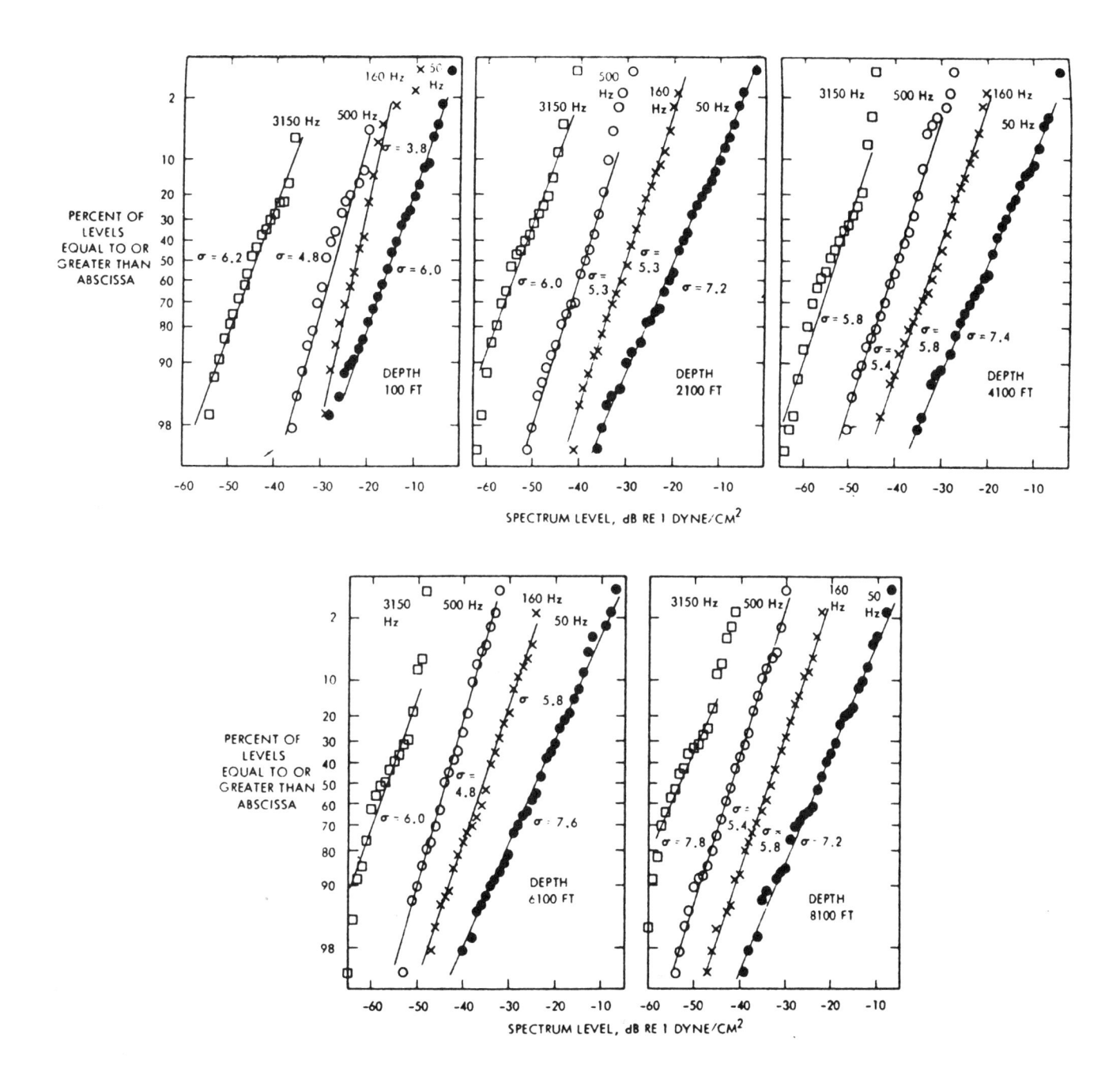

Figure 3-2. Cumulative distribution curves of 114 1½-minute samples taken at hourly intervals over a 2½-day period at five depths north of St. Croix. The values of σ are based on the straight lines drawn through the plotted points. Reference 40.

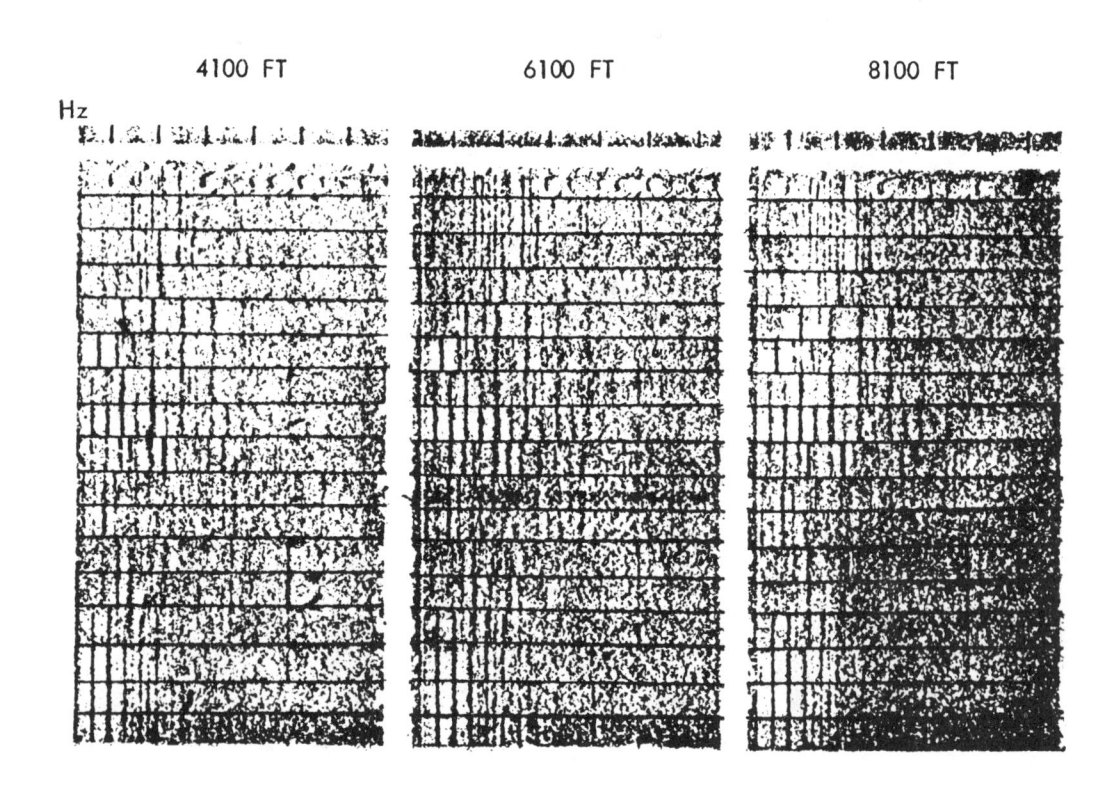

Figure 3-3. Sound spectrograms at three depths obtained hourly north of St. Croix. The frequency band for each spectrogram is 10 to 150 Hz. The harmonic series evident on most of the traces is the blade-rate line series of passing ships. Reference 40.

hourly intervals as observed during the just-mentioned overnight period at St. Croix. In the 10 to 150 Hz range covered by the sound spectrograms having a 1-Hz resolution, the distribution of line components evidently varies from hour to hour as the distribution of ship traffic varies. The regular harmonic series evident in many of the hourly samples is the "blade-rate series" of line components (with a fundamental of about 7 Hz, equal to the product of the shaft rotation rate and number of propeller blades) which has been estimated (83) to lie in the range 6.7 to 10.0 Hz for all of the merchant vessels of the world. This series of line components forms the dominant feature of the low-frequency spectrum of merchant ships--whose propellers regularly cavitate strongly at their normal operating speeds.

Much smaller variability would be expected in the frequency region of shipping noise in remote areas where distant ship traffic is sparse. Thus, over a 9-1/4 day period at a site identified only as in the southern North Atlantic Ocean, Hecht and Mole (84) found a value of σ of only 1-1/2 dB for 90°/o of both 10-sec and 10-min samples. The remaining 10°/o of the samples were higher and non-Gaussian and represented occasional passing local ship traffic at the site.

The noise due to shipping varies more rapidly with time than the noise due to the wind. This was demonstrated by Perrone and King (85) by analysis of data collected by Perrone at Bermuda and the Grand Banks. The autocorrelation time (time for the autocorrelation function to fall to zero) was 26 to 40 hours for wind noise, but only 4 to 8 hours for shipping noise.

Such hour-to-hour variability at different depths causes the noise profiles at a site to vary from hour to hour. Thus, the variability with depth is different at different times (see Figure 3-4), some depths being more

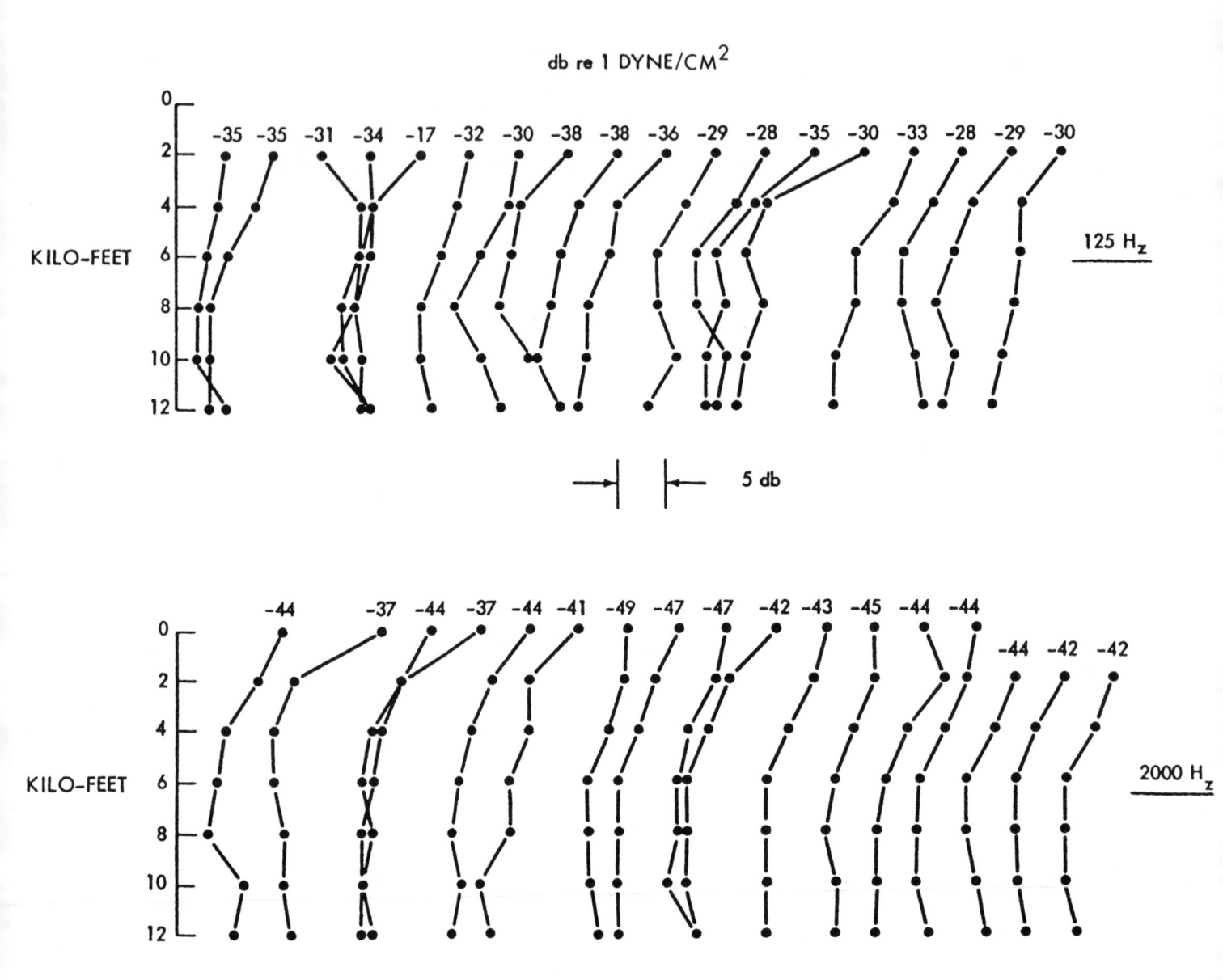

Figure 3-4. Variability of noise profiles from hour to hour. Each profile shows the average at different depths of 1½-minute noise samples. The number at the top of each profile shows the level at 2000 feet in dB relative to 1 dyne/cm². Reference 82.

noisy than others for short periods of time. At 125 Hz, such a variability represents the variability in the hourly distribution of ship traffic.

When tonal components are present, the variability of the noise depends upon the number of tones, or lines, occurring in the bandwidth of the analyzer. This number may depend upon the number of distant ships contributing to the noise, as well as upon the number of independently varying lines in the spectrum of a single ship within the analysis bandwidth. For a single steady tone, it has been shown by Urick (80) that the fluctuation of the intensity observed at a distant point in the sea has a Rician-distribution, with the fraction of the received signal that is random as a parameter. This "randomicity" is caused by scattering or multipath propagation, or both. When completely random (as a long-range signal is known to be), the Rayleigh distribution with σ = 5.6 dB occurs as a limit. But when, for example, four lines of equal level are present, σ reduces to 2.3 dB (86). A sequel paper to the foregoing permits numerical results to be obtained for the statistics of deep-sea shipping noise for more general cases (87). Tonal components in the oscillations of the Arctic ice cover, observed with seismometers, have been reported (88).

3.2.3 Long-Period Variations

Observations of noise made over one-year periods by means of hydrophones at Bermuda by Wenz (89) and Perrone (90) have been most valuable in throwing light on the long-term variability of noise at a single location. Perrone employed five bottomed hydrophones at various depths from 55 to 4500 m, and took 2-minute noise samples every 2 hours over a period of 1 year. No seasonal dependence of noise was found, except for that caused by different wind speeds in different seasons. None was found for ship noise in the range 17 to 112 Hz. Wenz employed two hydrophones, one north and one southwest of

Bermuda, over a 1-year period. Here also, a seasonal dependence amounting to 5 to 10 dB in 1/3-octave bands between 100 and 316 Hz, associated with the lower wind speeds in summer, was observed. This seasonal variability associated with wind speed is illustrated in Figure 3-5. In both studies just cited, the 20-Hz pulses (Section 7.3) produced by whales during the winter months (November through March) were a prominent seasonal characteristic of the low-frequency noise. The Wenz paper (89) compared the levels of Bermuda with previously observed data along the Pacific coast over a 2-year period, and found the two locations to be approximately comparable in level in the region 40 to 300 Hz where shipping noise is prevalent.

Some seasonal noise dependence may also be expected from seasonal changes in sound transmission, such as the existence of a thick mixed-layer duct in winter and a thin one, or its absence altogether, in summer. A difference in sound transmission was suggested by Walkinshaw (91) to account for a 7-dB range in mean noise level between a winter maximum and a summer minimum observed at the Bahamas and at Bermuda; however, the seasonal variability of wind speed and shipping cannot be discounted in these observations. Over a one-year period of measurement of noise with a bottomed hydrophone on the Grand Banks, Piggott (67) observed 3.5 dB more noise at all frequencies in winter, under isovelocity conditions, than in summer, when a negative velocity gradient, with its poor propagation because of downward refraction, existed.

Because of an increasing number of ships and increases in ship speed, ship size and propulsive power, Ross (92) has estimated that in the 25-year period 1950-1975, the ambient levels due to distant shipping increased by some 10 dB, so as to dominate the spectrum in some areas where shipping noise was previously insignificant. Much of this noise has doubtless been the result of the emergence and prevalence of supertankers in many shipping lanes. Although

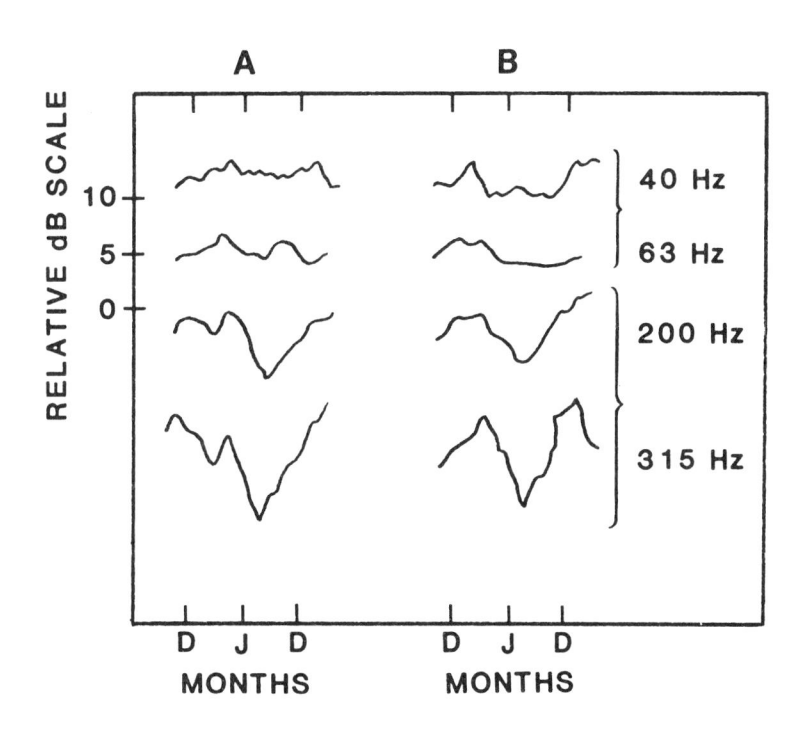

Figure 3-5. Monthly average ambient noise levels at two locations near Bermuda. A, north of Bermuda; B, southwest of Bermuda. The months are indicated by D, December; J, June. Reference 89.

in recent years the number of very large ships at sea has diminished, the resulting quieting of the sea may have been compensated for by an increase in the number and size of fishing boats, as well as by an increase in off-shore oil exploration and drilling.

3.3 Noise Variability in Shallow Water

Shallow water is noteworthy for its variability. Waters close to shore and in busy harbors are dynamic locations where many and rapid noise changes take place. One such location is off Fort Lauderdale, Florida, where a variety of pleasure boat traffic, biologic activity, and weather changes in the form of squalls and thunderstorms occur. A contrasting location is one 10 miles offshore in the Gulf of Maine, where ship traffic noise and biologic noise are rare and weather changes were slight during the time noise recordings were obtained. Figure 3-6 shows hourly samples of noise at these two locations over a 24-hour period, observed by Urick (93). The greater hour-to-hour variability and higher average noise level at the Florida site is evident.

Although shallow water is notably characterized by variability as a result of a highly variable background of ship and biological activity, the level at high frequencies and high wind speeds, when wind noise is prevalent, is remarkably constant from site to site at the same wind speed. Figure 3-7 is a compilation of measured levels at 1000 Hz taken from the literature at a wide variety of locations under "quiet" conditions, when ship and biological noises were absent and only the nondescript noise characteristic of the wind could be heard by listening. The levels reported in the literature are remarkably concordant, not only among themselves but with the "Knudsen" curves that have been considered for many years to be representative for both deep

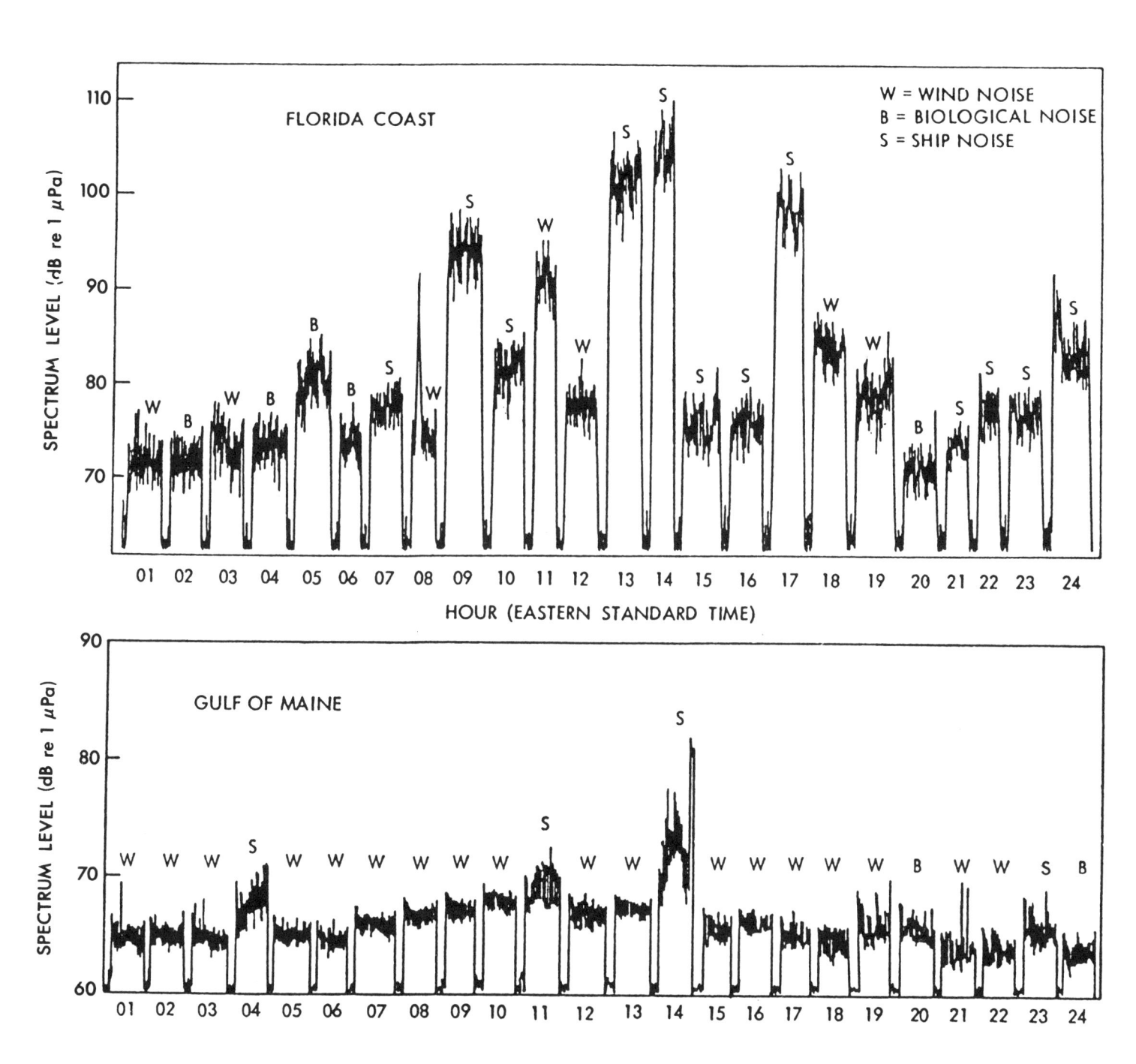

Figure 3-6. Playouts of 1-minute noise samples taken at hourly intervals at two locations off the east coast of the U.S. Frequency band 100 to 200 Hz. The letters give the type of noise as identified by listening to the samples. Reference 93.

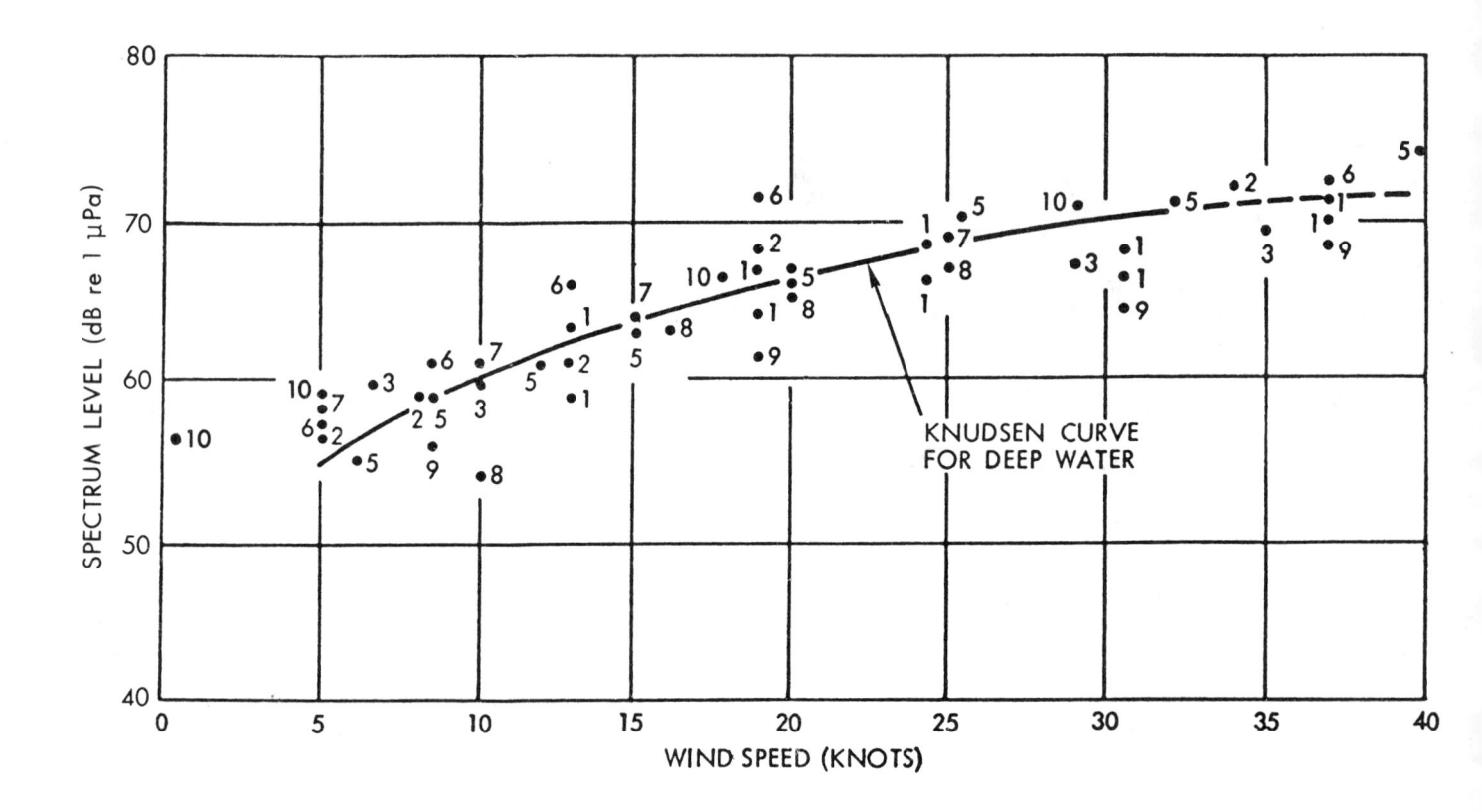

Figure 3-7. 1000 Hz noise levels at various shallow water locations obtained from the literature. The numbers alongside each plotted point denote the report or paper from which the value was obtained. The solid curve is not an average through the points but is a plot against wind speed of values from the Knudsen spectra at 1000 Hz. Reference 93.

3-14

CONDITIONS FOR FIGURE 3-7

Reference	Location	Water Depth	Hydrophone Depth
1	Off Perranporth, England	120 ft	Bottomed
2	Bermuda	1200 ft	Bottomed
3	Long Island Sound	26-52 fms	10-179 ft
4	Various	Deep Sea	Shallow
5	Scotian Shelf	120 & 168 ft	Bottomed
6	Five Pacific Ocean Locations	Not Stated	Not Stated
7	Florida Coast	140 & 530 ft	Bottomed
8	Gulf of Maine	360 ft	Bottomed
9	Bermuda	180 ft	Bottomed
10	Russian Coast	100-130 m	Bottomed

REFERENCES

1. D. Smith and G. Wearden, Measurement of Sea Noise Level and Wind Velocity at A.E.S. Perranporth, June 1957 - Nov 1957, ARL/N 21/L, June 1958, unpublished.

2. W.C. Beckman, Low Frequency Ambient Sea Noise, JUA 5, July 1955, unpublished.

3. H.R. Johnson, Underwater Sound Transmission in Shallow Water, Part I: Ambient Noise Measurements, WHOI Ref. 52-48, 1952, unpublished.

4. V.O. Knudsen, R.S. Alford and J.W. Emling, Underwater Ambient Noise, Journal of Marine Research 7, 410, 1948, Fig. 4.

5. C.L. Piggott, Ambient Sea Noise at Low Frequencies in Shallow Water of the Scotian Shelf, JASA 36, 2152, 1964, Fig. 3.

6. G.M. Wenz, Acoustic Ambient Noise in the Ocean: Spectra and Sources, JASA 34, 1936, 1962. Average of five Pacific Ocean areas, Fig. 1.

7. R.J. Urick, Ambient Acoustic Environment of Shallow Water Off Ft. Lauderdale, FL, NOLTR 70-72, 1970. Site I, 2000 Hz levels increased by 5 dB.

8. R.J. Urick, The Underwater Acoustic Environment at Two Contrasting Shallow Water Locations, Naval Ordnance Laboratory TR 71-69, 1971.

9. E.M. Arase and T. Arase, Ambient Sea Noise in the Deep and Shallow Ocean, JASA 42, 73, 1967, Fig. 8. 700 Hz data decreased by 3 dB.

10. V.I. Bardyshev, A.M. Velikanov and S.G. Gershman, Experimental Studies of Underwater Noise in the Ocean, SPA 16, 512, 1971.

and shallow water. The reason for this unusual degree of consistency of the data is doubtless the fact that wind noise originates at the sea surface immediately over the measuring hydrophone, together with a high bottom loss at the 1000-Hz frequency.

Resorting to theory, a simple analysis of the sound field produced by a dense, plane distribution of random dipoles each radiating with an intensity proportional to the square of the cosine of the grazing angle (m = 1) shows that (1) half the noise comes from an area on the surface of radius equal to the hydrophone depth, and (2) the intensity is independent of measurement depth in the absence of attenuation and bottom effects. In passing, we should note that the compiled levels of Figure 3-7 increase at the rate of 6 dB per wind speed doubled, meaning that the intensity of wind noise increases as the square of the wind speed.

The variability of shallow-water noise was also observed for two four-week periods at two nearby sites off Fort Lauderdale, Florida in 34 and 150 m of water by Urick and Bradley (94). The distribution of the 500 hourly levels each one minute long at the two sites was found to be distinctly non-normal, with an excessive occurrence of high levels, even at frequencies as high as 4000 Hz, due to nearby passing pleasure and shipping craft. Excluding 20°/o of the noisy samples, the remaining 80°/o had a standard deviation of about 6 dB, tending to be higher at low frequencies than at high. The levels at both sites tended to show a diurnal variability, with ship noise greater by day than by night, and with biologic noise the reverse, occurring more strongly at night than during the day.

In shallow water, propagation sometimes affects the ambient background, though in a different way than it does distant shipping noise in deep water. For example, in the Bristol Channel separating England and Wales in 39 m of water, a series of narrowband tonal components was found by Weston (95) to occur in the noise background at frequencies of 10, 28 and 47 Hz. These frequencies varied with the water depth during the tidal cycle, with each frequency corresponding to the cut-off frequency of one of the normal modes of sound propagation. That the tidal cycle can sometimes strongly affect sound propagation in shallow water was clearly shown in one exeriment (96), where a 3-foot range of tide in 60 feet of water caused a 15-dB change of the level from a source about a mile away as a result of interference between the two lowest, and therefore strongest, normal modes of transmission.

Closer to shore, the noise of surf has a different character (97). In the surf zone less than 60 meters from a rocky, gravelly beach, the noise sounded like a series of "clicks," presumably made by pebbles colliding as they were agitated by waves; as a wave rolled upon shore, the number of "clicks" and the noise level were observed to increase abruptly. Further out, at a distance of 20 km from the shoreline, the noise distribution was found to become Gaussian over periods of 20 minutes or less.

An interesting kind of variability was observed by Wenz (98) with bottomed hydrophones at six locations in water depths between 150 and 3000 feet. He found a diurnal variability of 1.5 to 5.0 dB that occurred at midnight local time regularly throughout the year; at one station daily changes of 10 to 20 dB occurred during the period of the summer solstice (at maximum solar declination) but no changes occurred at the time of the winter solstice. Strangely, the maxima occurred at the same local time each day (at midnight) and did not follow the tidal cycle. The cause or causes of these diurnal variations remain a mystery; many possibilities were examined by Wenz, ranging from biological to even extra-terrestrial causes, but all were rejected as being unlikely.

3.4 Correlation Between Frequency Bands and the Effect of Beamwidth

Most studies of variability have dealt with the statistics of variability in particular frequency bands. On the other hand, one study, made by Nichols and Sayer (99), has been concerned with the correlation of ambient noise fluctuation <u>between</u> narrow frequency bands. In this study, a total of 572 consecutive samples, each 640 sec long, were correlated between different 6-Hz frequency bands over the range 5 to 150 Hz. The total recording time (t_3) was thus 4.3 days; all tonal CW signals that might originate from discrete ships were excluded from the data. The observations were made at an unspecified site with a water depth of 2150 m in the North Atlantic during the month of February. Between two frequency bands, f_1 and f_2, the results showed the correlation coefficient of the intensity averaged over 640 sec to be (by definition) 1.0 when $f_1 = f_2$, but to fall to about 0.5 for a ratio of one octave between f_1 and f_2. Between any two bands, the coefficient was seldom less than 0.3. The results were unfortunately complicated by the admixture of ship noise and wind noise throughout the frequency band in the stormy month of the year in which the data were obtained. Examples of the cross-correlation between 50 Hz and 100 Hz and other frequency bands are shown in Figure 3-8.

Using a high-gain narrow beam array, Anderson (100) found the fluctuation of the ambient background to be higher at the output of the beamformer than at the output of a single element of the array. That is to say, the envelope fluctuation of the background had an entirely different character when observed with a highly directional receiver than with a single element. With directional beams 1.5° x 6° wide at 3 1/2 kHz, envelope spectrum levels were found to be 20 to 30 dB higher than those predicted for a Gaussian process. Figure 3-9 shows comparative fluctuation spectra for an essentially

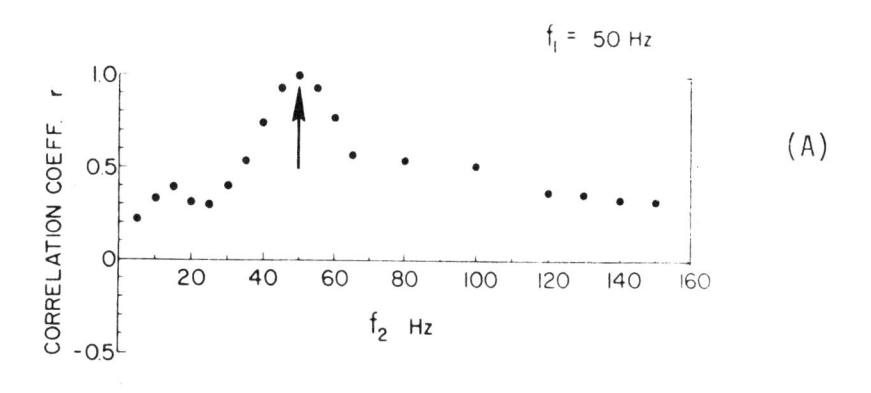

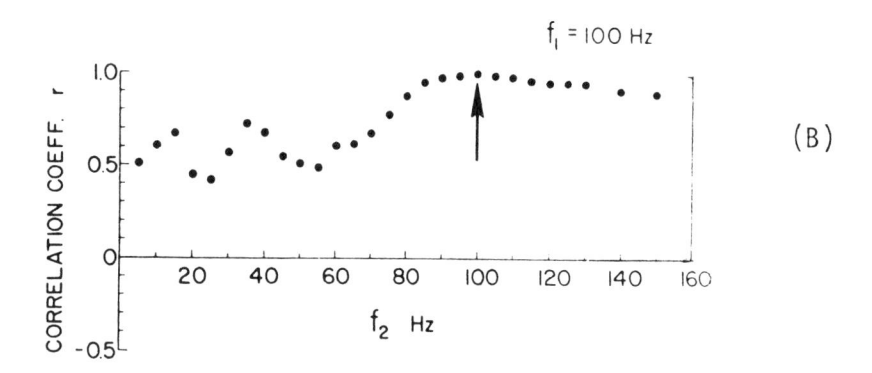

Figure 3-8. Correlation coefficients of 10.7-minute samples of noise between different 6 Hz frequency bands. In (A) the 50 Hz band was correlated with other bands. In (B) the 100 Hz band was correlated with other bands. Reference 99.

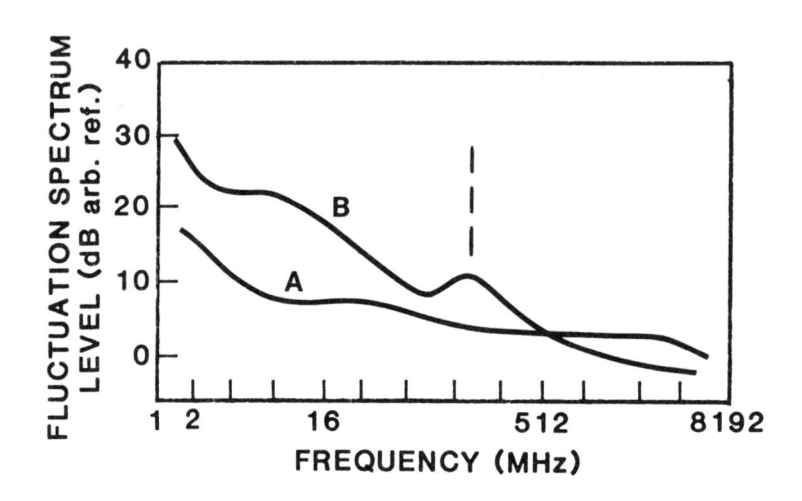

Figure 3-9. Relative fluctuation spectra. Curve A, nearly non-directional transducer; curve B, in a beam 1.5° x 6° wide looking upward. The frequency band is 800 to 1600 Hz. The dashed line shows the ocean spectrum peak due to swell. Reference 100.

non-directional transducer (A) and for a narrow beam (B) looking upward and receiving largely surface noise.

Such noise fluctuation in narrow beams has an effect on signal detectability. Although it is well known that doubling the integration time of a processor increases the detectability of steady sinusoidal signals by 1.5 dB in a background of steady noise, it has been shown theoretically (101) that this applies only for very short (<0.5 sec) or very long (>120 sec) integration times; in between, the improvement of detectability can be as small as 0.06 dB per time-doubled. When the signal fluctuates as well as the noise, the effect of an output averager or integrator is probably even more degraded; however, this aspect of the subject appears not to have been investigated in the literature.

Concerning the stationarity of ambient noise, Arase and Arase (102) found that in 1/3-octave bands the noise was stationary for less than 3 minutes at frequencies between 100 and 1600 Hz. On the other hand, Jobst and Adams (103), analyzing the noise between 75 and 260 Hz in narrow bandwidths between 0.129 and 0.448 Hz, found stationarity for periods up to 22 minutes. The discrepancy between the two studies may perhaps be attributed to the different frequencies and bandwidths used, but remains for the most part a mystery.

3.5 Summary

It is self-evident to say that the ambient background at some particular location varies with time because of the time variability of the sources of noise. At a single site, variations of ship traffic and of wind speed over a period of time cause variations of noise level. It seems impossible to predict the variability--in terms of the standard deviation of an ensemble of noise samples--except by comparison with data already obtained at a similar location. Such data is sparse. There is a suggestion in the data that the standard deviation of noise samples roughly equals 6 dB for shipping noise when dealing with one-minute samples over periods of a few hours or days; such a figure, however, is highly uncertain and doubtless represents only an order of magnitude. Wind-noise variability is less easily predictable, unless meteorological observations are available at a nearby land station, from which a reasonably reliable prediction of level and its variability can presumably be made.

The subject of noise variability is one that has not been systematically studied in its own right, but in the literature seems to have been only incidental to other research objectives.

4. DEPENDENCE ON RECEIVER DEPTH

4.1 Introduction

One would expect almost intuitively that ambient noise originating at the sea surface should decrease with depth, if only because of the attenuation due to absorption. This is indeed the case; however, theory shows that the decrease is less than might be anticipated, as a result of the small absorption coefficient of sound in sea water. On the other hand, low-frequency shipping noise originating at a distance is not appreciably affected by absorption because its frequency is low; its dependence upon depth is governed by the many other factors affecting long-range propagation in the sea.

4.2 Theory

4.2.1 Surface Noise

Let the sea surface be represented by an assemblage of random, closely packed radiators distributed over the plane sea surface. It was shown long ago by Urick (104) and later by Bradner and Parvulescu (105) that if the intensity radiated by an element of area dA at unit distance at an angle θ with the vertical is taken to be of the form

$$I(\theta) = I_0 \cos^{2m}\theta$$

then, at depth h, where h is in meters, the intensity received from the entire surface is

$$I(h) = 2\pi \cdot (\alpha h)^{2m} \cdot I_0 \cdot \int_{\alpha h}^{\infty} \frac{e^{-y}}{y^{2m+1}} \, dy,$$

where α is the absorption coefficient in units of reciprocal meters, with the simplifying assumptions of (1) straight-line paths (no refraction), (2) an infinite water depth (no bottom reflection) and (3) an α independent of depth. For the most reasonable source assumption, that of dipole radiation corresponding to m = 1, the above integral can be integrated by parts to give

$$I(h) = \pi \cdot (\alpha h)^2 \cdot I_0 \cdot \left[E_0(\alpha h) + e^{-\alpha h}\left((\alpha h)^{-2} - (\alpha h)^{-1}\right) \right]$$

where $E_0(\alpha h)$ is the exponential integral $\int_{\alpha h}^{\infty} \frac{e^{-y}}{y} dy$ for which tables are available (106). Figure 4-1 gives the calculated attenuation of surface noise with depth for a number of frequencies; the results would not be appreciably different for m = 0 or m = 1/2. We note that high-frequency surface noise is rapidly attenuated and does not penetrate to great depths, and at moderate depths becomes so low in level that it is overcome by the noise of molecular bombardment and often (for practical purposes) by the electronic noise of a pre-amplifier, if one is used.

At shallow depths, where αh is very small, I(h) becomes independent of depth. It can be shown (104) as a curiosity that for m = 1 the intensity I_0 of a unit area of the sea surface at unit distance in the vertical is 5 dB less than I(h); also, half the noise intensity comes from a circular area of sea surface of radius equal to h, while the other half comes from the sea surface beyond h.

4.2.2 Shipping Noise

Several theoretical papers have considered the depth distribution of long-distance shipping noise (107), (108), (109). The general finding is that there should be little depth variation for a typical velocity profile as long as the receiver remains within the Deep Sound Channel. But below the channel (i.e., below the critical depth, defined as the depth at which the velocity is the same as at the surface), the noise level should decrease rapidly with depth as the trapped modes of wave theory attenuate with increasing distance

4-2

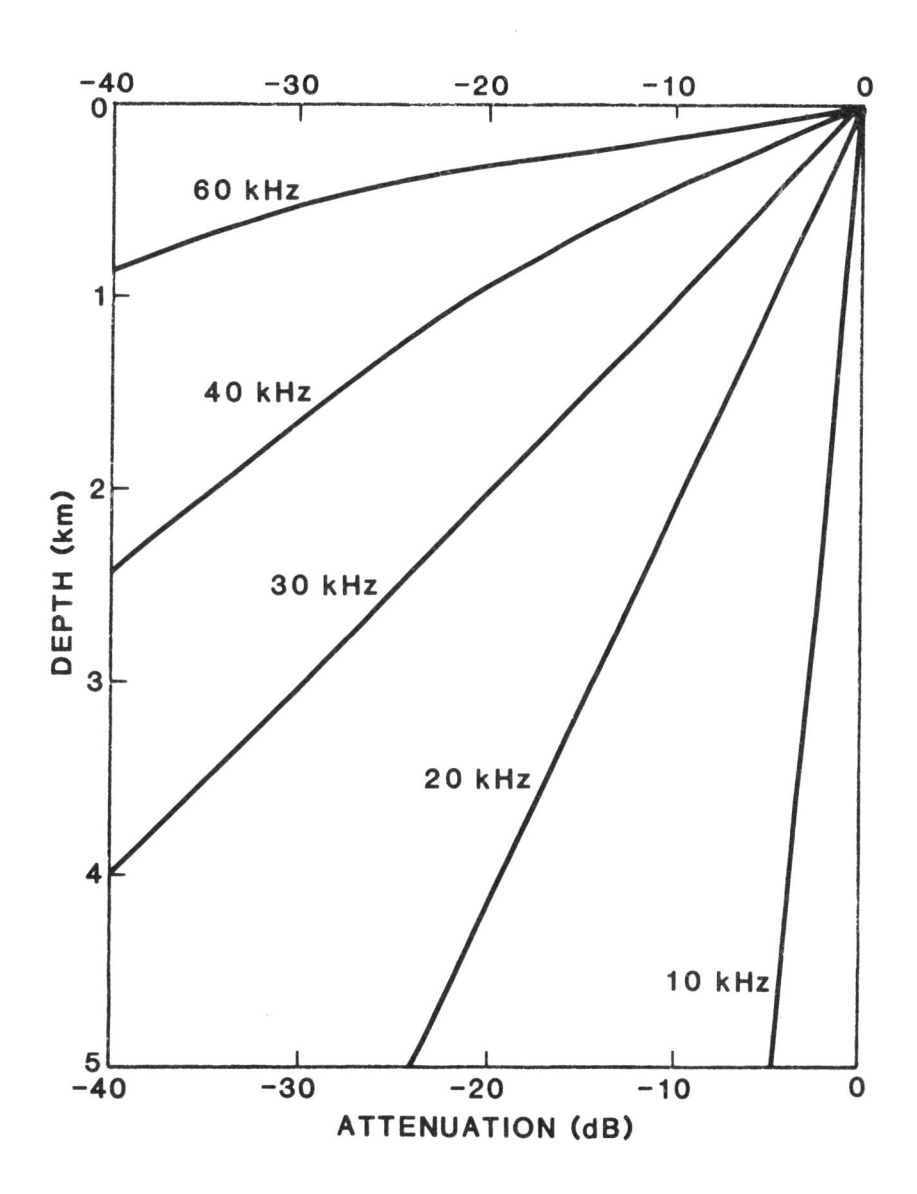

Figure 4-1. Theoretical attenuation of surface-generated noise with depth caused by absorption, assuming straight-line paths, an infinite water depth and a constant absorption coefficient with depth.

below the channel (110). Also, close to an absorbing bottom the modes will interact with the bottom and further attenuate the noise (111). This latter effect is overriding for velocity profiles having a limited depth excess--those with the water depth only slightly greater than the critical depth. Another factor causing the noise level to be low near the bottom is the bathymetry in the vicinity of the receiver, causing shadowing by intervening bottom ridges and seamounts. All these effects cause a deep hydrophone near the sea bed to be more quiet than one near the surface or at mid-depth.

4.3 Field Observations

Three instrumental techniques have been used to determine the variation of noise with depth in deep water. In one--the earliest--the noise picked up by bottomed hydrophones at different depths in the same general area was measured and compared. However, this method does not reveal the true depth profile that exists in water of constant depth since the propagation to the different hydrophones from distant sources is not the same. Another method, yielding simultaneous data, uses a string of hydrophones located at different depths along a single supporting conducting cable. A third method uses a depth controllable programmable hydrophone package yielding noise recordings made at the different pre-selected depths. Finally, a fourth method is to use unmanned freely drifting probes. Two of these were called SONODIVER, which could hover at a preset depth while recording noise, and SPARBUOY, which transmitted the noise from a shallow hydrophone to a mother ship. A description of these interesting vehicles and some preliminary results may be found in Reference 112.

Over the years a number of hydrophones have been installed on the bottom at different locations in different water depths south and west of Bermuda.

Perrone (113) compared the noise at different sites in different water depths and reported the results as depth profiles. The plots showed a strong positive apparent noise gradient (i.e., level increasing with depth) at low frequencies and a strong negative apparent gradient at high frequencies. Figure 4-2 shows an example of Perrone's findings for a wind speed of 20 knots. These profiles, at variance with subsequent measurements in deep water of constant depth, doubtless show the effects of long-range propagation of ship noise to hydrophones located on the bottom in different depths of water. Subsequently, Arase and Arase (70) compared a 30-fathom bottomed hydrophone at Bermuda with another bottomed hydrophone 3.3 miles away at a depth of 900 fathoms. As mentioned previously (Section 2.8), the shallow unit showed noise levels lower than those of the deep unit by 2 to 16 dB, depending on wind speed and frequency, and was wind-speed dependent at all frequencies. The sloping bottom between the two receivers evidently screened the shallow hydrophone from the ship noise arriving at the deep hydrophone.

As long ago as 1957 the manned bathyscaphe TRIESTE made four dives at two sites in the Tyrrhenian Sea off the west coast of Italy down to 1000 and 3200 meters, measuring the noise en route. The resulting profiles obtained by Lomask and Frasetto (114) showed no depth dependence in sea state zero in the 1000-meter area in the frequency range 10 to 240 Hz, but showed a strong negative noise gradient in sea state 2 in the 3200-meter area. The two sets of data were obtained in different areas at different times.

Simultaneous measurements at a single site can be made by means of hydrophones suspended from a cable and recording simultaneously. Figure 4-3 illustrates one such arrangement that was used by Urick, Lund and Tulko (82) at a deep-water site on the side of a deep basin 20 x 120 km in size north of the island of St. Croix. The prevailing velocity profile is given at the

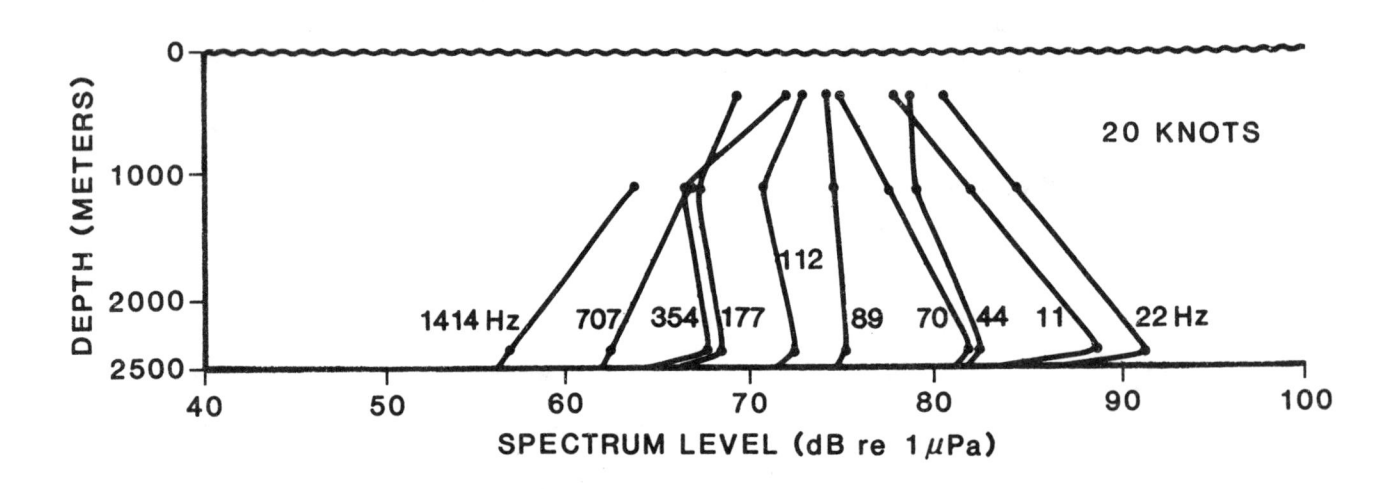

Figure 4-2. Noise profiles at bottomed hydrophones off Bermuda. Reference 113.

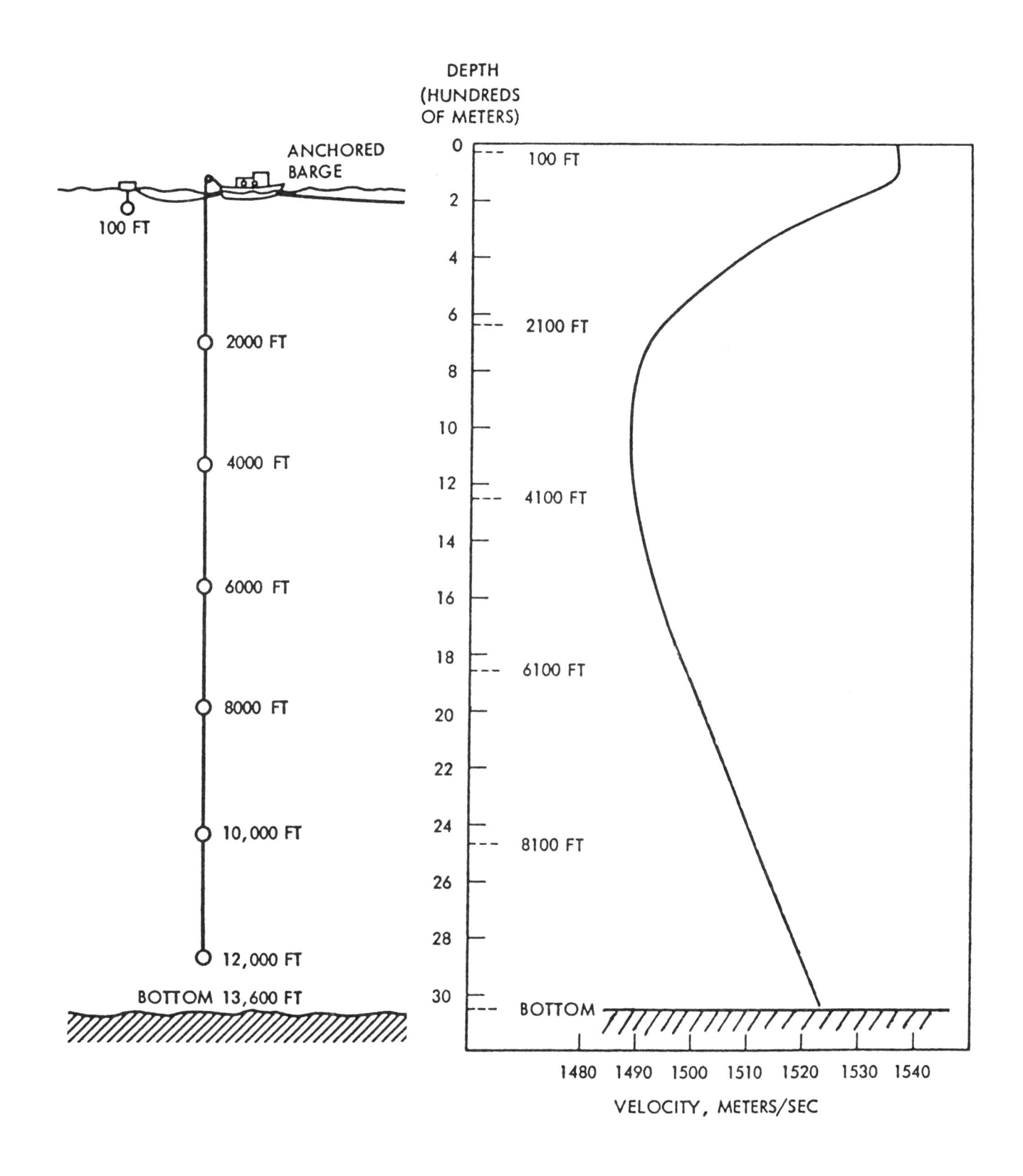

Figure 4-3. Hydrophone string and velocity profile for measurements of the depth profile north of St. Croix. The velocity profile shows no critical depth. Reference 82.

right. Figure 4-4 shows average levels of 18 consecutive 1-1/2-minute hourly noise samples taken at depths 2000 feet apart. The profiles show a gradual decrease of level with depth, with a sharp decrease from 100 to 2000 feet, probably as a result of the greater amount of sound within the surface duct (Section 4.4), together with a small amount of contamination by noise of the suspension system and the barge at the 100-ft hydrophone. The noise gradient tends to be greater at low frequencies than at high. We note from Figure 4-3 that there was no critical depth in the velocity profile.

Essentially similar results were subsequently obtained by Morris (64) in the open ocean of the northeastern Pacific with hydrophones dangled from the research platform FLIP (Floating Laboratory Instrument Platform). Figure 4-5 shows for this experiment the velocity profile, hydrophone depths and average noise profiles in 1/3-octave bands at different frequencies. As at St. Croix, there is a decrease of noise with depth at low frequencies with a smaller, or a complete absence of, a decrease with depth at 500 Hz as wind noise overcomes the dominance of ship noise. Below the critical depth, the fall-off with depth is steeper as the bottom is approached as the result of a stripping away of progressively more energy of refracted sound in the Deep Sound Channel as the hydrophone descends below the critical depth. This quieting below the critical depth was also observed by Kibblewhite, Ellis and Hampton (115), also in the northeast Pacific Ocean, using hydrophones moored up from the bottom at three locations and at three depths -- on the channel axis, at the critical depth and near the bottom. Figure 4-6 is an example of the profiles obtained from 10-second noise samples averaged over a 24-hour observation period. At an entirely different location--the approaches to the English Channel--a moored hydrophone string was used by King (33). Between depths of 206 m and 2764 m, above the critical depth, no depth variation and no wind-speed

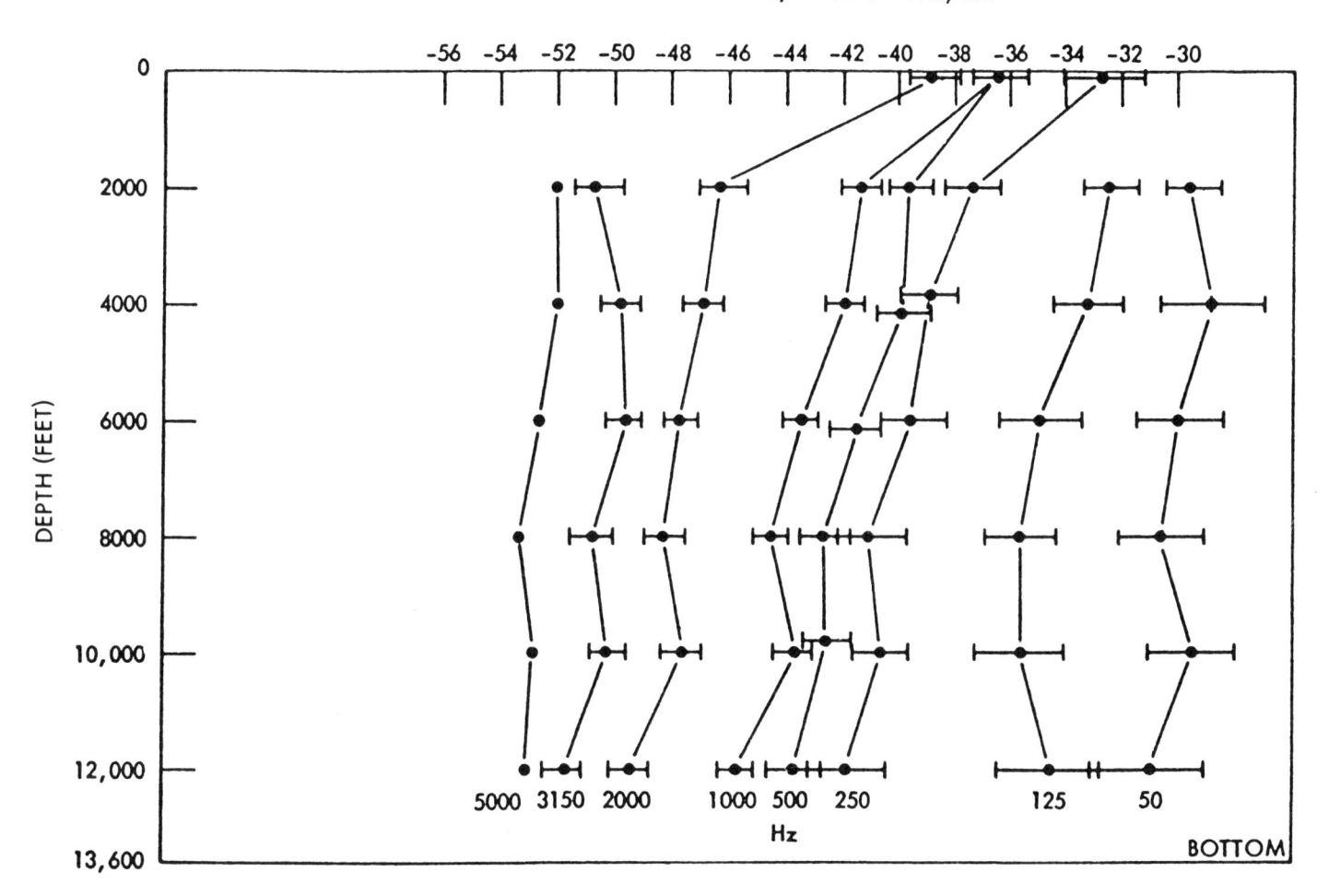

Figure 4-4. Depth profile of noise north of St. Croix measured in 1/3 octave frequency bands. The horizontal lines at each depth show one standard deviation of 18 consecutive hourly noise samples each 1½-minutes long. Reference 82.

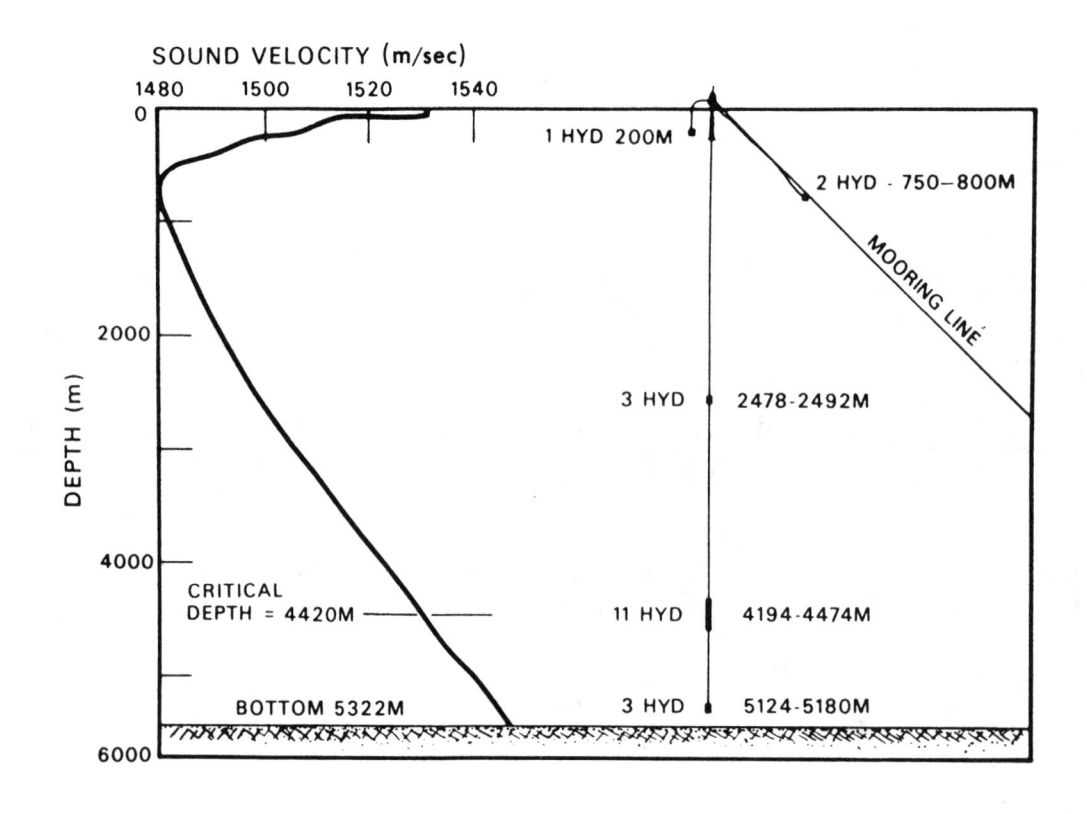

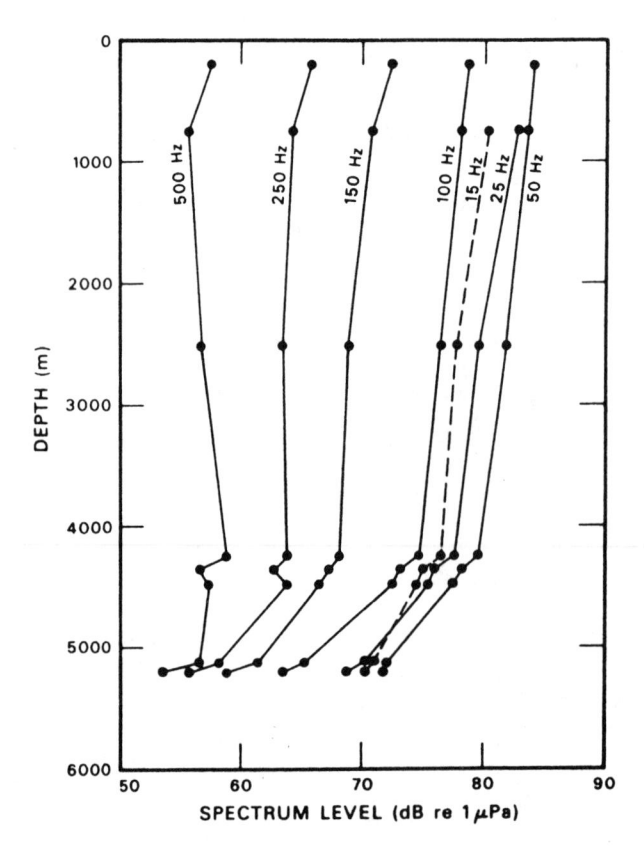

Figure 4-5. Velocity profile, hydrophone depths and depth profiles measured in the northeast Pacific Ocean. Reference 64.

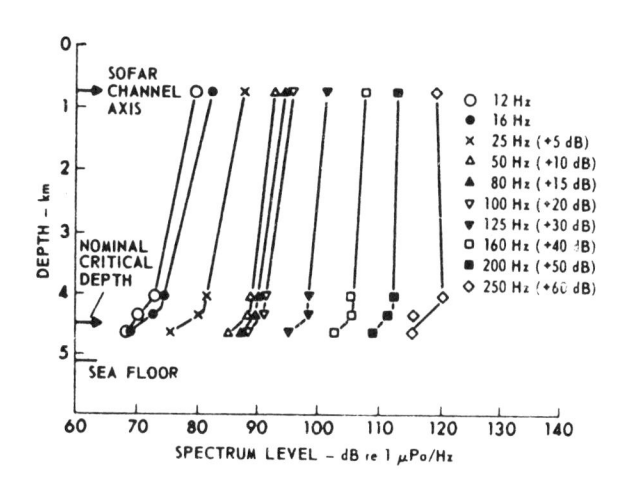

Figure 4-6. Depth profiles at and below the Deep Sound Channel axis in the northeast Pacific Ocean. Reference 115.

variation could be found in this heavily trafficked measurement area over the frequency range 31.6 to 3160 Hz.

4.4 Noise in the Surface Duct

There is more noise in the surface duct (called the mixed layer by oceanographers), then at depths below the duct (116). This is shown by Figure 4-7 giving a comparison of noise spectra obtained with two sonobuoy hydrophones at depths of 60 and 300 ft in a 120-foot layer. Above the lowest trapped frequency in the duct, as given by a formula adapted from radio literature (117, p. 151), there is 5 to 10 dB more noise in the duct than below it. Correlograms show (see Figure 6-8) that this noise is likely to originate in distant areas of sea surface and to travel nearly horizontally via surface duct paths to the receiver. Thus, the surface duct, when it is a good duct for the transmission of signals, is apt to be more noisy as well.

4.5 Effect of Bathymetry

The bottom topography at considerable distances can have an appreciable effect on the noise level and depth variation of the ambient background. For example, at two locations 30 miles apart near Greenland, depth profiles obtained with the AUTOBUOY programmable self-contained free-diving recording system were appreciably different (118). At one site the bathymetry was open to the east; at the other, the Reykjanes Ridge formed a barrier to the east extending upward to about 1000 meters from the surface. The second site was found to be about 10 dB more quiet at deep depths and low frequencies than the first site and therefore to have a more steeply sloping noise profile.

The bathymetry can enhance the received noise level as well as reduce it. Morris (119) found that a passing supertanker was received more strongly,

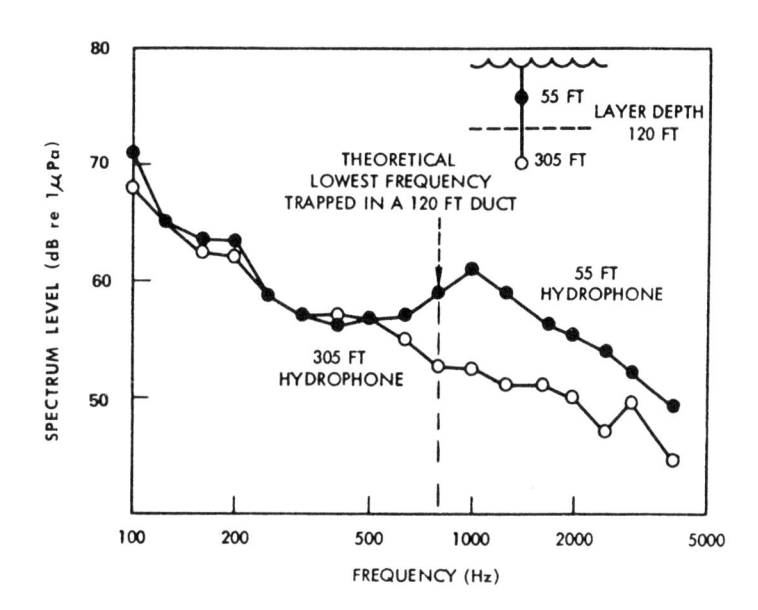

Figure 4-7. Ambient noise spectra within and below a surface duct. Reference 116.

by as much as 10 dB, when it passed near a seamount at a distance of 15 miles, and by up to 5 dB when it crossed the 1000-fathom contour of the continental slope. The former enhancement was surmised to be the result of reflection from the seamount; the latter was ascribed to the "down-slope conversion" effect (see Figure 2-20), wherein the sound from a shallow source is "converted" by the bottom slope into sound propagating in the Deep Sound Channel. No such increases were noted on hydrophones below the critical depth.

Enhancement of noise may also be produced by a range-dependent velocity profile, such as that across the Gulf Stream between the slope water to the northwest and the Sargasso Sea to the southeast. In all likelihood, because of the changing profile, a deep hydrophone at Bermuda receives part of its ambient background from coastal ship traffic passing several hundreds of miles to the west and north of Bermuda.

4.6 Summary

From the evidence available at the present time the variation of noise level with depth in deep water may be summarized as follows. At frequencies where ship noise is dominant, there is a gradual quieting with depth amounting to a few decibels, down to the critical depth (if there is one in the profile), below which the quieting beomes more rapid. The quietest depth is at or near the ocean floor. Sea surface noise at higher frequencies tends to have no variation with depth because this noise is of local origin, except for a greater amount of noise in the surface duct, if one is present. At frequencies above 10 kHz, however, theory shows a strong negative gradient of surface noise with depth due to absorption--so much so as to make surface noise negligible at high kilohertz frequencies at moderate depths.

5. DIRECTIONALITY OF AMBIENT NOISE

5.1 Introduction

For a number of years after World War II, nothing was known about the directionality of ambient noise. During this early period, all measurements were made with nondirectional hydrophones, which of course reveal nothing about directionality. The noise field of the sea was tacitly considered to be isotropic--that is, uniform in all directions--even though it was recognized that since the noise level in the frequency range of the Knudsen curves depended strongly on sea state or wind force, the ambient background must originate in some way at the sea surface and, therefore, should produce a vertically directional sound field.

5.2 Theoretical Considerations

In order to obtain a theoretical expression for the directionality of surface noise, let us consider the simplest imaginable model. Referring to Figure 5-1, let the sea surface consist of random densely packed noise radiators each radiating according to $I(\theta) = I_0 \cos^n \theta$, where $I(\theta)$ is the intensity of a unit area of radiators at unit distance at angle θ from the vertical, and I_0 is the intensity in the downward direction ($\theta = 0°$). At a point P at depth h, the intensity from a ring of radiators of area dA at angle θ is

$$dI = \frac{I(\theta)}{\ell^2} dA = \frac{I(\theta)}{\ell^2} 2\pi r \, dr$$

on the assumption of straight-line paths, no attenuation, and infinite water depth (no bottom reflection). Expressing r, ℓ and dr in terms of h and θ, we

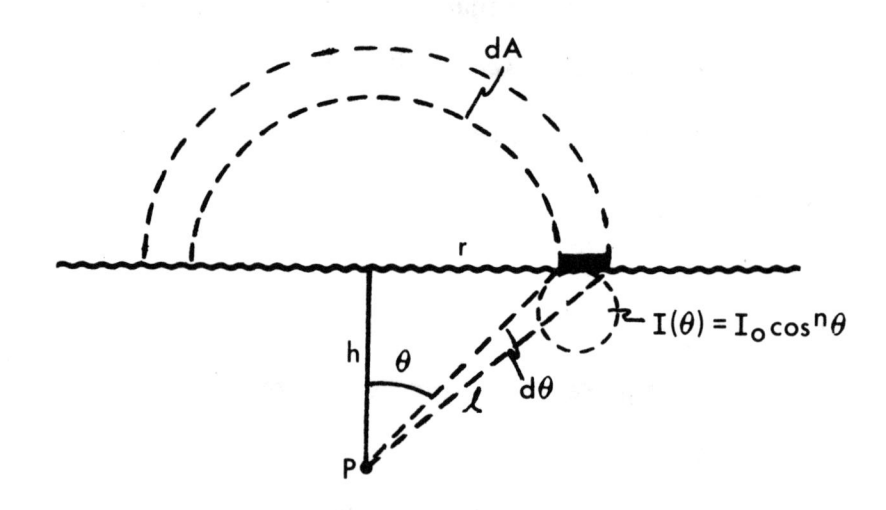

Figure 5-1. Geometry of a simple model for the vertical directionality of noise from random sources on a plane surface, each radiating like $\cos^n \theta$.

get
$$r = h \tan\theta; \quad \ell = h \sec\theta; \quad dr = h \sec^2\theta \, d\theta$$
so that
$$dI = 2\pi I_0 \cos^n\theta \tan\theta \, d\theta$$

The intensity per unit solid angle ψ thus becomes

$$\frac{dI}{d\psi} = \frac{2\pi I_0 \cos^n\theta \tan\theta \, d\theta}{2\pi \sin\theta \, d\theta} = I_0 \cos^{n-1}\theta$$

Concerning n, the most plausible assumption is that the radiating elements of the sea surface behave like small dipoles, in which case n = 2. For this case we have, for the intensity of surface noise per unit solid angle at angle θ,

$$\frac{dI}{d\psi} = I_0 \cos\theta$$

This result has been plotted in Figure 5-5 to show the comparison with the measured directional patterns of Axelrod, Schoomer and Von Winkle. A more realistic model including attenuation and bottom reflection losses has been obtained by Talham (120).

The problem of deriving the directional pattern of the ambient noise background from the measured outputs of a finite number of directional beams requires the solution of a series of integral equations. For a vertical array, let $F(\theta,\theta_0)$ represent the beam pattern of the array when steered at angle θ_0 to the vertical, and let $N(\theta)$ be the noise intensity per unit solid angle at angle θ. Then, if the noise is horizontally isotropic, the voltage output $V(\theta_0)$ of the θ_0 beam is given by

$$V^2(\theta_0) = 2\pi R \int_{-\frac{\pi}{2}}^{\frac{\pi}{2}} F(\theta,\theta_0) N(\theta) \sin\theta \, d\theta,$$

where R is a conversion factor representing the response of the array. From the measured beam outputs $V^2(\theta_o)$, one for each beam, the best possible evaluation of the function $N(\theta)$ needs to be found. If there are k beams, the result is a set of k simultaneous equations for finding $N(\theta)$.

This problem has been solved in a number of ways. One, used by Fox (121), is by successive approximations. Another is to solve the k equations by inverting the coefficient matrix (122) and by finding a least-squares approximation to the noise field (123). Still another is by spatial harmonic analysis (124). In actuality there is no single solution for $N(\theta)$; there is a "principal solution" that oscillates least and is the most physically plausible representation of the true noise field. Finding this principal solution has been the subject of a paper by Anderson and Tittle (125). An alternate approach to the problem is through the cross-spectrum or cross-correlation between pairs of elements of the array, as suggested by Yen (126).

5.3 Vertical Directionality

To measure the vertical directionality of noise in the sea, an array of some kind, capable of forming directional beams, is required. The first, and at the same time, the most elaborate and sophisticated array employed for this purpose, was described in 1958 by Anderson (127) and was used at sea by Becken (122) as well as by Rudnick and Squier (128). This was a 3-dimensional array of 32 small hydrophones that was called "the great stellated icosohedron" because of its size, symmetry and star-shaped appearance. Thirty-two beams could be formed in different oblique directions. With it, measurements were made by Becken of both the horizontal and vertical directionality of the noise background in deep water at several depths down to 1000 ft in the octave band 750 to 1500 Hz. Figure 5-2 shows directional patterns obtained at sea states

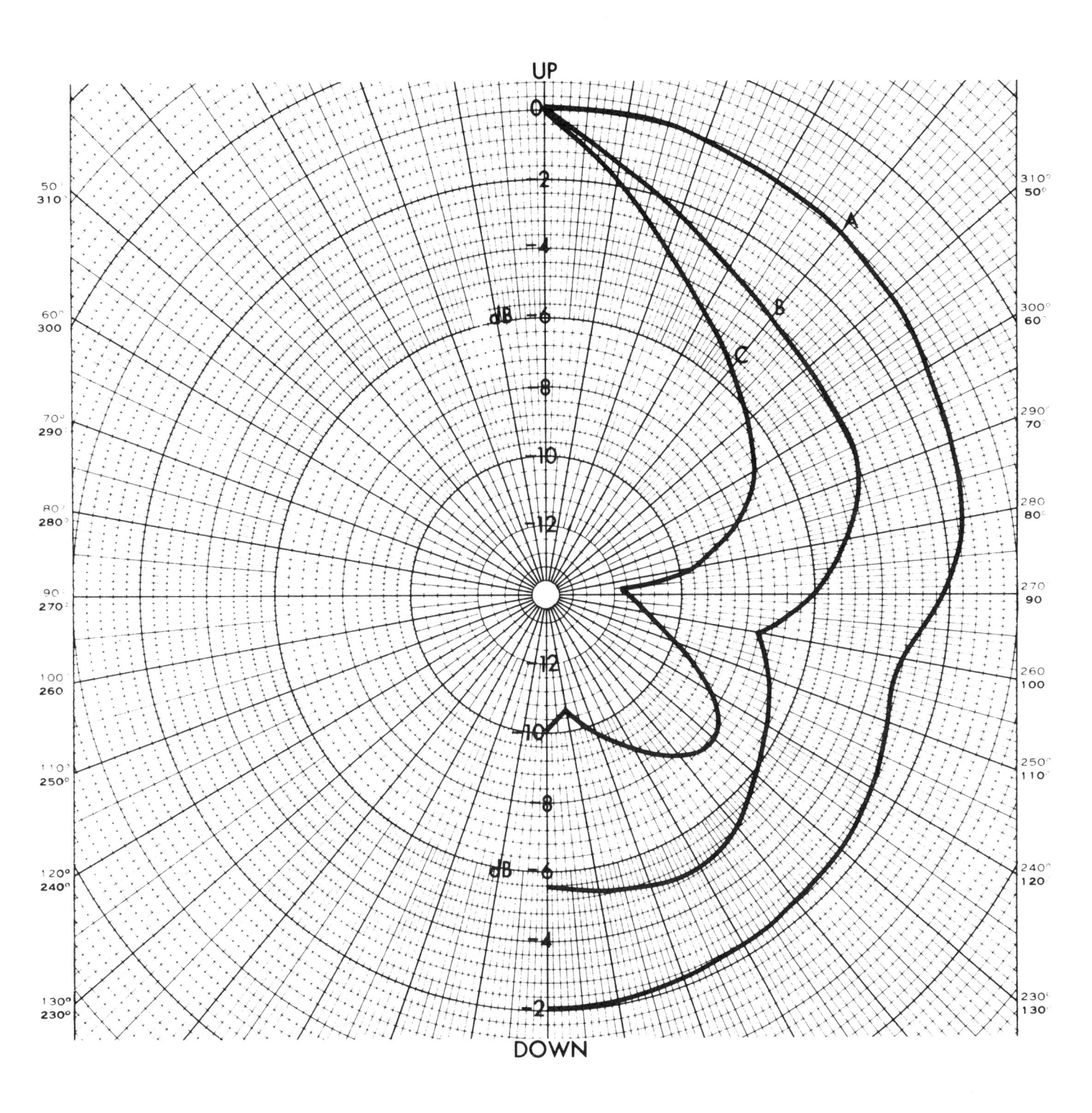

Figure 5-2. Vertical directional patterns at three sea states in the band from 750 to 1500 Hz as obtained by Becken (122) with a 3-dimensional array. A - sea state ½, depth 1000 ft; B - sea state 1½, depth 130 ft; C - sea state 3, depth 560 ft.

1/2, 1-1/2 and 3. With increasing sea state the patterns are seen to become more directional because of the increasing proportion of surface noise relative to shipping noise. Subsequent measurements with the Anderson-Becken array, supported from the large spar buoy FLIP were reported by Rudnick and Squier (128) to be largely in agreement with the earlier results of Becken (122).

A vertical string of hydrophones forms a simple array for the determination of vertical directionality. One such array was a bottom-anchored array called FAB (Fixed Acoustic Buoy) consisting of 21 hydrophones 16 ft in overall length buoyed up from the bottom by a float. Included within the bottom anchor were the analog delay lines and the stepping switches required for beam formation. Figure 5-3 shows three patterns obtained by Forster (129) with this system for three sea states in a 1/2-octave band centered at 2000 Hz. Again we see an increasing vertical directionality with increasing sea state.

Considerably longer vertical arrays are required to investigate lower frequencies. One of these, called the Trident Vertical Array (TVA), has had considerable use in a variety of ambient noise and propagation studies. This was a vertical string of forty unequally spaced hydrophones 300 ft long buoyed up by 60 ft to the lowest hydrophone from the bottom in 14,400 ft of water (see Figure 6-4). Each hydrophone was connected to shore at Bermuda 25 miles away by a multi-conductor cable. The TVA as used by Fox (121) had a total of 84 beams covering the upper hemisphere over the frequency range 112 to 1414 Hz. No downward beams were used, presumably because the proximity of the array to the bottom would make the results meaningless. Figure 5-4 is a sampling of the patterns obtained. The decibel scale here is absolute (dB re 1µPa per Hz per steradian of solid angle). In these results more noise

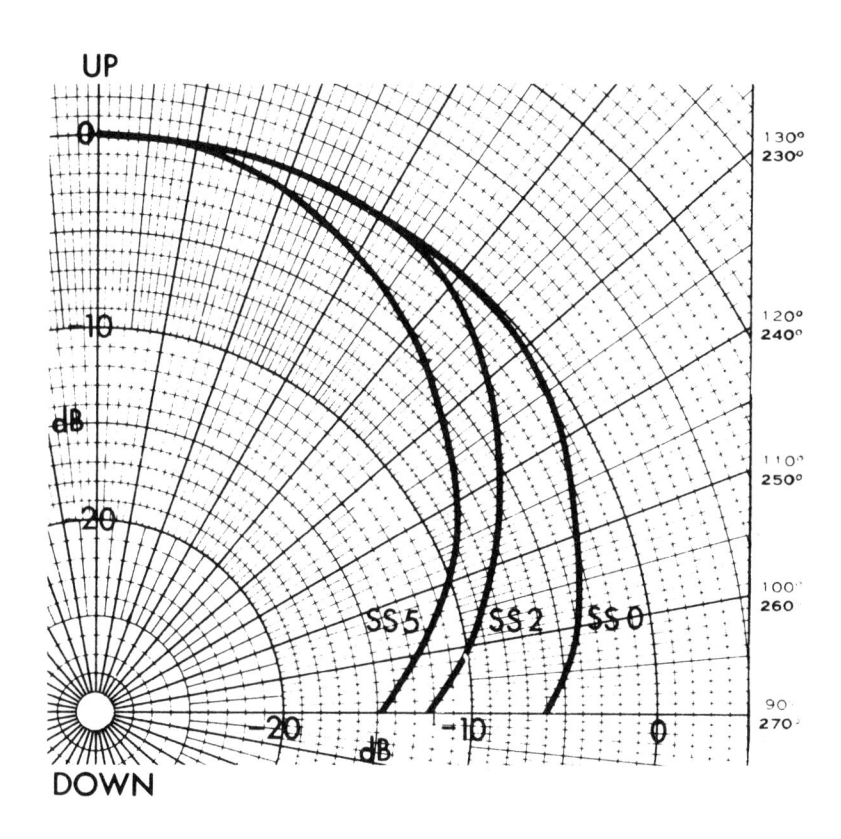

Figure 5-3. Vertical directional patterns as obtained by Forster in a 1/2-octave band centered at 2000 Hz. Reference 129.

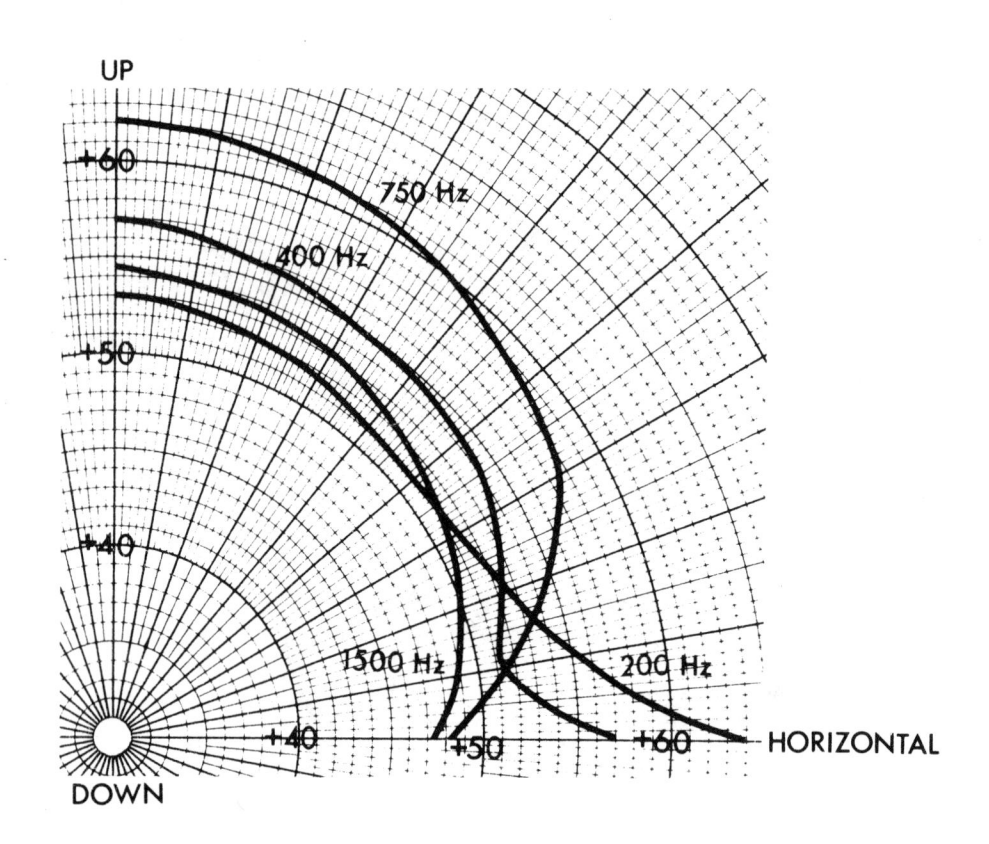

Figure 5-4. Vertical directional patterns in 1/3-octave bands at four frequencies in sea state 3. The scale is spectrum level in dB re 1 µPa per steradian. Reference 121.

comes from the horizontal at 200 and 400 Hz than from the vertical--a characteristic of low frequencies where, as in this area, ship noise dominates surface noise.

This reversal of the vertical directional pattern was confirmed by subsequent measurements with the TVA by Axelrod, Schoomer and VonWinkle (124). Figure 5-5 shows patterns at 112 and 891 Hz with wind speed as a parameter; with increasing wind speed, the patterns become less directional at 112 Hz and more directional at 891 Hz as the "mix" of the two types of noise changes with wind speed.

Another example of low-frequency noise directionality is given in Figure 5-6, as measured by a long vertical line array at a depth of several hundred feet (130). At 90 Hz there is 30 dB more noise per unit solid angle arriving from the horizontal than from the vertical; at 380 Hz the difference is less, due to the admixture of surface noise; at a still higher frequency the noise would doubtless be sensibly isotropic.

More recent data have been obtained by Burgess and Kewley (32) by means of a vertical string of 31 hydrophones 180 m long suspended at a depth of 300 m at four locations between Australia and New Zealand. Examples of the directional patterns otained are given in Figure 5-7. On comparing the two wind speeds and the two frequencies in this figure, we note once again the increasing dominance of wind noise with increasing wind speed and increasing frequency. Surprisingly, in the measurements at 37, 70, 140, 280 and 560 Hz a wind-speed dependence was found in all frequency bands, although the dependence was slight in the 37- and 70-Hz bands at 8 knots and above. In the higher bands the level increased with wind speed at the rate of about 4.5 dB per wind speed doubled, or as 15 times the logarithm of the wind speed.

At intermediate depths in deep water, a self-contained free-floating 12-element array called AUTOBUOY has proven useful. This array could be

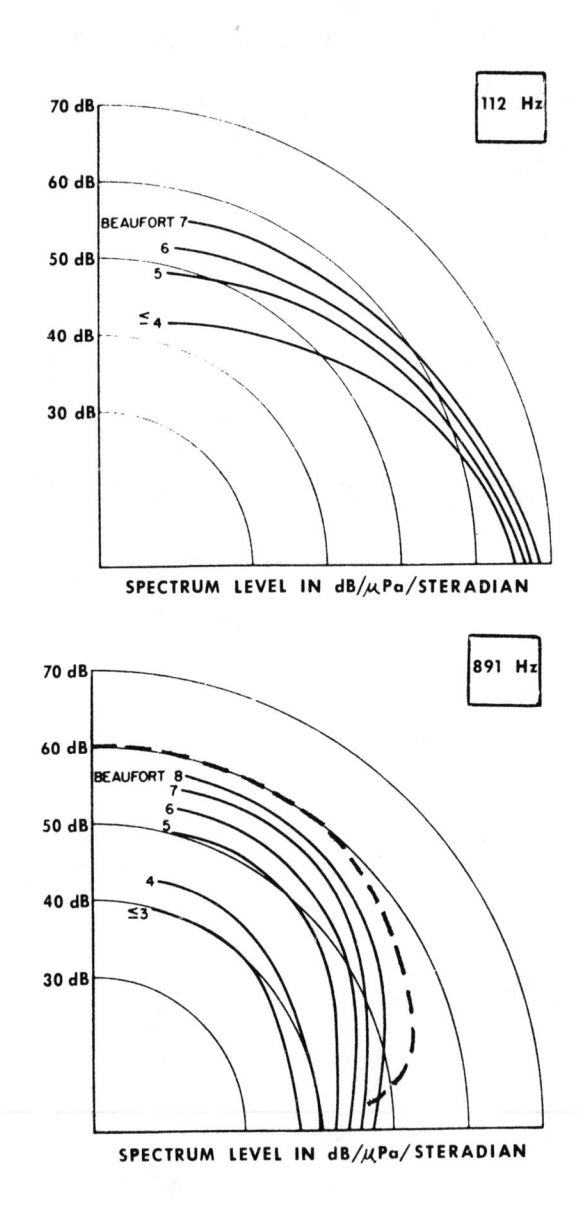

Figure 5-5. Ambient noise directional patterns at 112 and 891 Hz as reported by Axelrod, Schoomer and Von Winkle (124). The dashed curve is a curve of 10 log cos θ that results from a simple model.

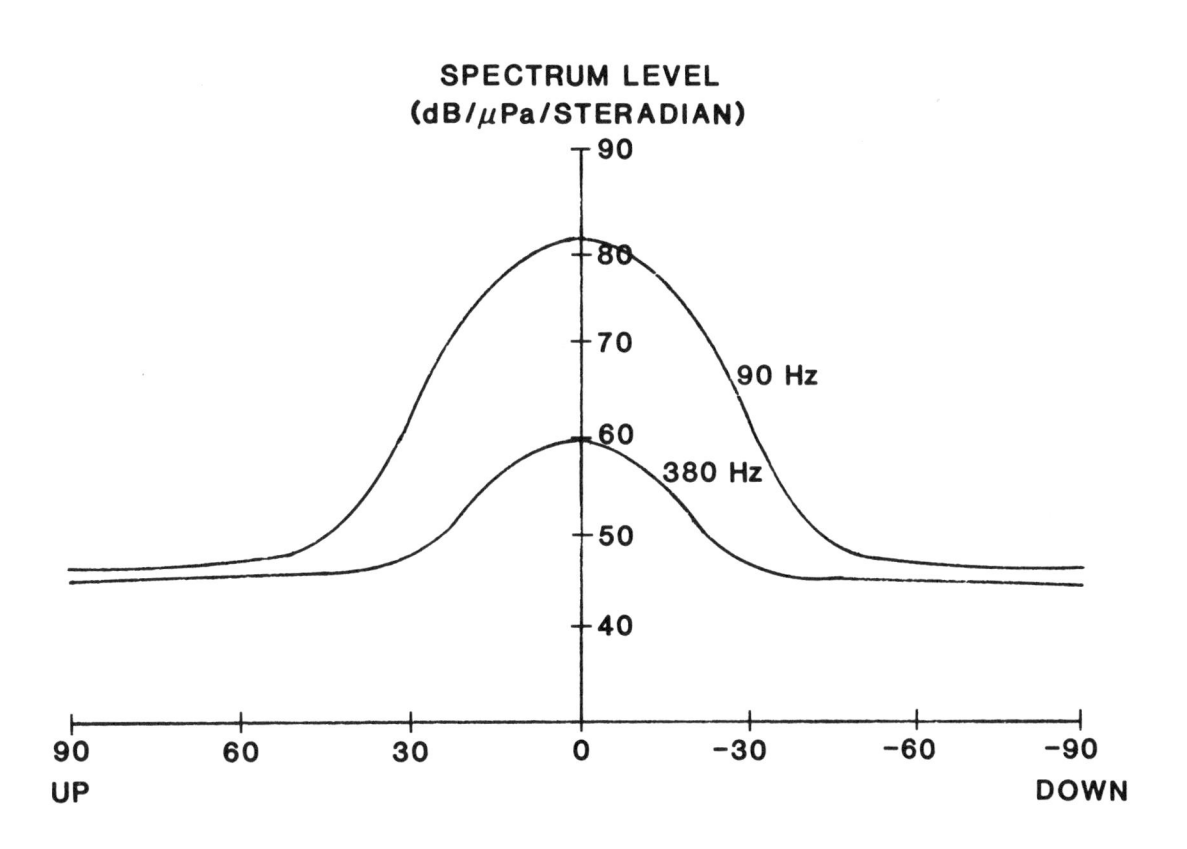

Figure 5-6. Directional patterns of ambient noise at two frequencies. Reference 130.

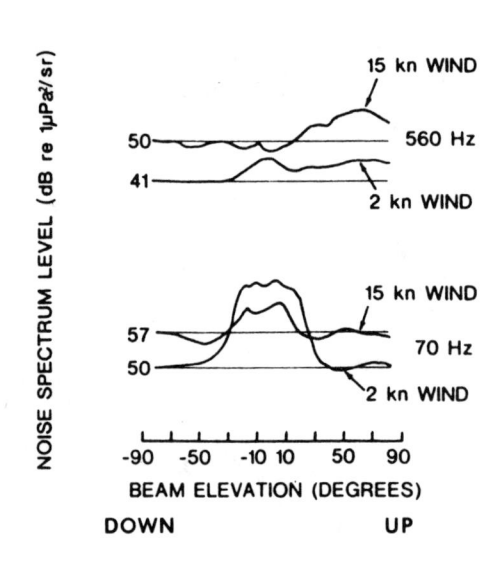

Figure 5-7. Ambient noise directionality measurements at two frequencies in the South Fiji Basin. Reference 32.

programmed to acquire and record data at different depths over an 8-hour
period. In measurements in the South Fiji Basin north of New Zealand, in an
area of light ship traffic, data was gathered at 300, 1200 (sound channel
axis), 2200 and 3100 meters by Browning and others (131). Figure 5-8 shows
some of the patterns obtained. In such areas of light ship traffic we cannot
ignore the possibility that surface noise originating in shallow water near
islands, ridges and seamounts becomes "converted" by down-slope propagation
(the "slope enhancement effect") to noise traveling with low loss in the Deep
Sound Channel; by this process, horizontally arriving noise may come to
exhibit a dependence upon wind speed that otherwise would not be expected to
occur.

In the deep sound channel with water of uniform depth and a constant
velocity profile, a ray diagram shows that sound rays from a distant shallow
source cannot reach a receiver on or near the axis of the channel (the depth
of minimum velocity) by nearly horizontal paths. The fact that shipping
noise is so highly collimated, or directional, along the horizontal is therefore
somewhat of a mystery. The most likely explanation, according to Dashen and
Munk (222), is the downslope conversion process, wherein the noise of coastal
ship traffic is bounced off the continental slope and floods the Deep Sound
Channel with nearly horizontal rays.

5.4 Variation of Vertical Directionality with Depth

Because of different propagation to different depths, the vertical
directionality of noise traveling within the Deep Sound Channel should vary,
to some extent, with depth. On the axis of the channel in deep water of
uniform depth, a ray diagram shows that sound from distant surface sources can
arrive only at angles of ± 12 to $15°$ to the horizontal; at smaller angles only

5-13

low-level diffracted, scattered or slope-enhanced energy can be received. Thus, the noise directional pattern on the axis should show two maxima at ±12 to 15° (above and below the horizontal) with a null in the center. Off-axis, at shallower or deeper depths, the double-humped pattern should contract and eventually become a single horizontal lobe at depths near the limits of the channel. The theory for this effect has been worked out by Cavanagh and Renner (109). Unfortunately, the double-humped characteristic may not be seen in actual data because of lack of resolution of the measurement array, or, more likely, because of the absence of the ideal conditions implied in the

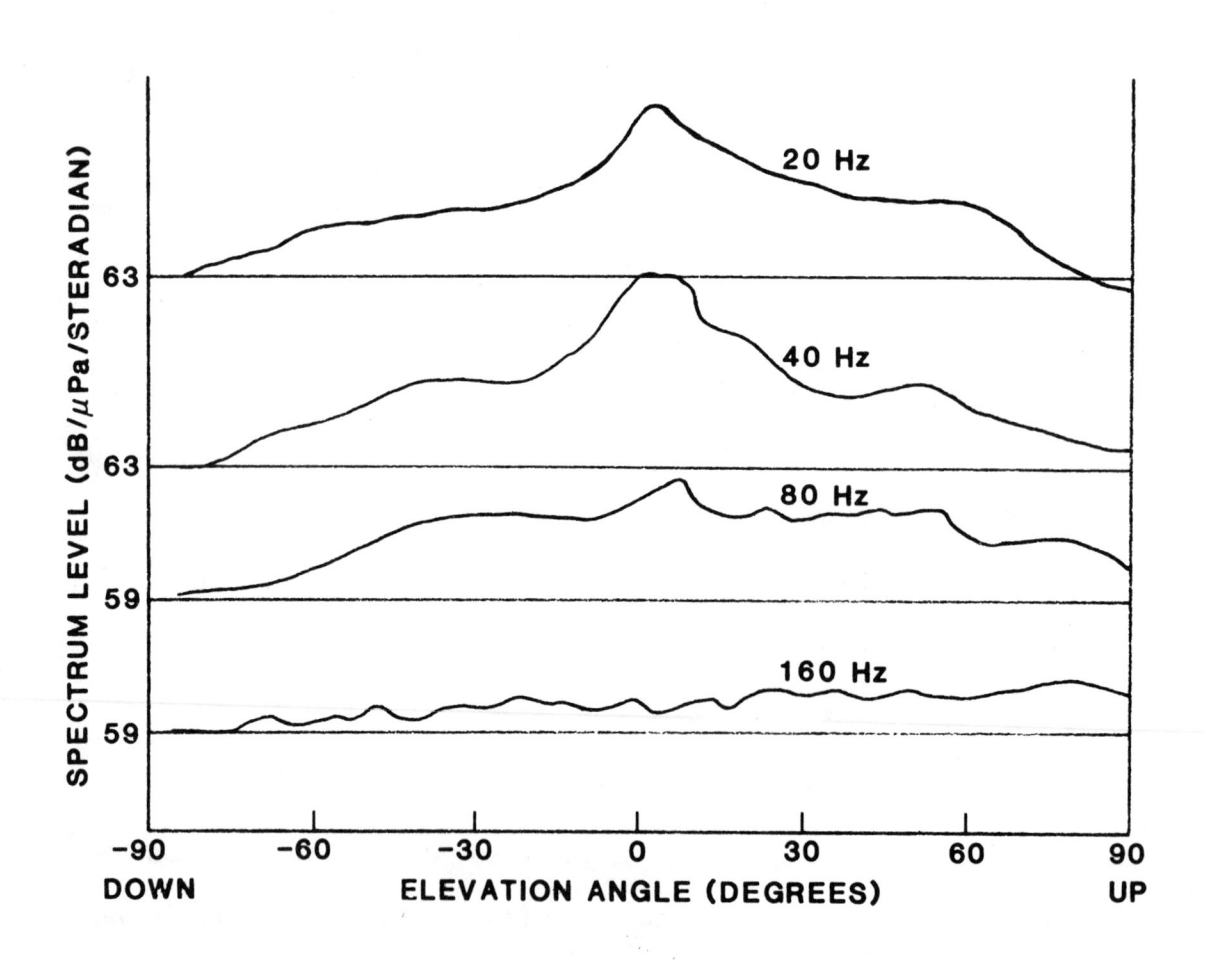

Figure 5-8. Vertical noise patterns on the Sound Channel axis (1200 m) in the South Fiji Basin. Reference 131.

simple theory. Anderson (132) suspended a vertical array of 20 equally spaced hydrophones (532 m long) from FLIP. Directional patterns were obtained in the North Pacific Basin at a number of depths between 713 m on the axis and 3781 m at the "critical depth" where the sound velocity was the same as at the surface. Figure 5-9 shows the patterns obtained at these two depths. While the pattern is more narrow at 3781 m than at 713 m, there is no sign of the ideal double-peaked pattern on the axis. The filling-in of the pattern near the horizontal was attributed to the shallowing of the velocity profile toward the north in the North Pacific. Ship noise originating in the dense traffic in the major shipping lanes toward the north travels down the channel along paths near the channel axis to the measurement location. Another mechanism might be "down-slope conversion" into the Deep Sound Channel of sound from ships on the continental slope of the Pacific Ocean basin.

5.5 Horizontal Directionality

The development of long, flexible towed line arrays for detection and surveillance has given importance to the directionality of ambient noise in the horizontal plane. However, compared to the vertical, only a relatively few unclassified investigations have been made. Becken (122), using the 32-beam volume array mentioned above, has obtained the only extant information on the horizontal directionality of sea surface noise. As shown by Figure 5-10, more noise was found to come from directions parallel to the crests and troughs of the waves and swell than at right angles. The effect was attributed to be shadowing of the sources of noise by the troughs of the intervening waves and swell in the perpendicular direction, whereas unimpeded transmission occurs in directions parallel to the waves and swell.

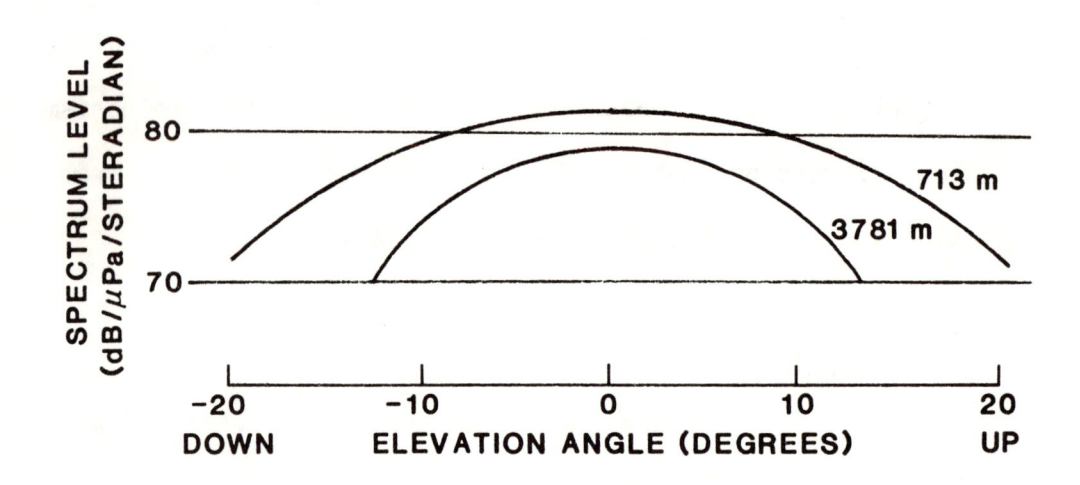

Figure 5-9. Vertical directional patterns at two depths at a frequency of 38 Hz. The 713-m depth was at the axis of the Deep Sound Channel; 3781 m was the critical depth. Reference 132.

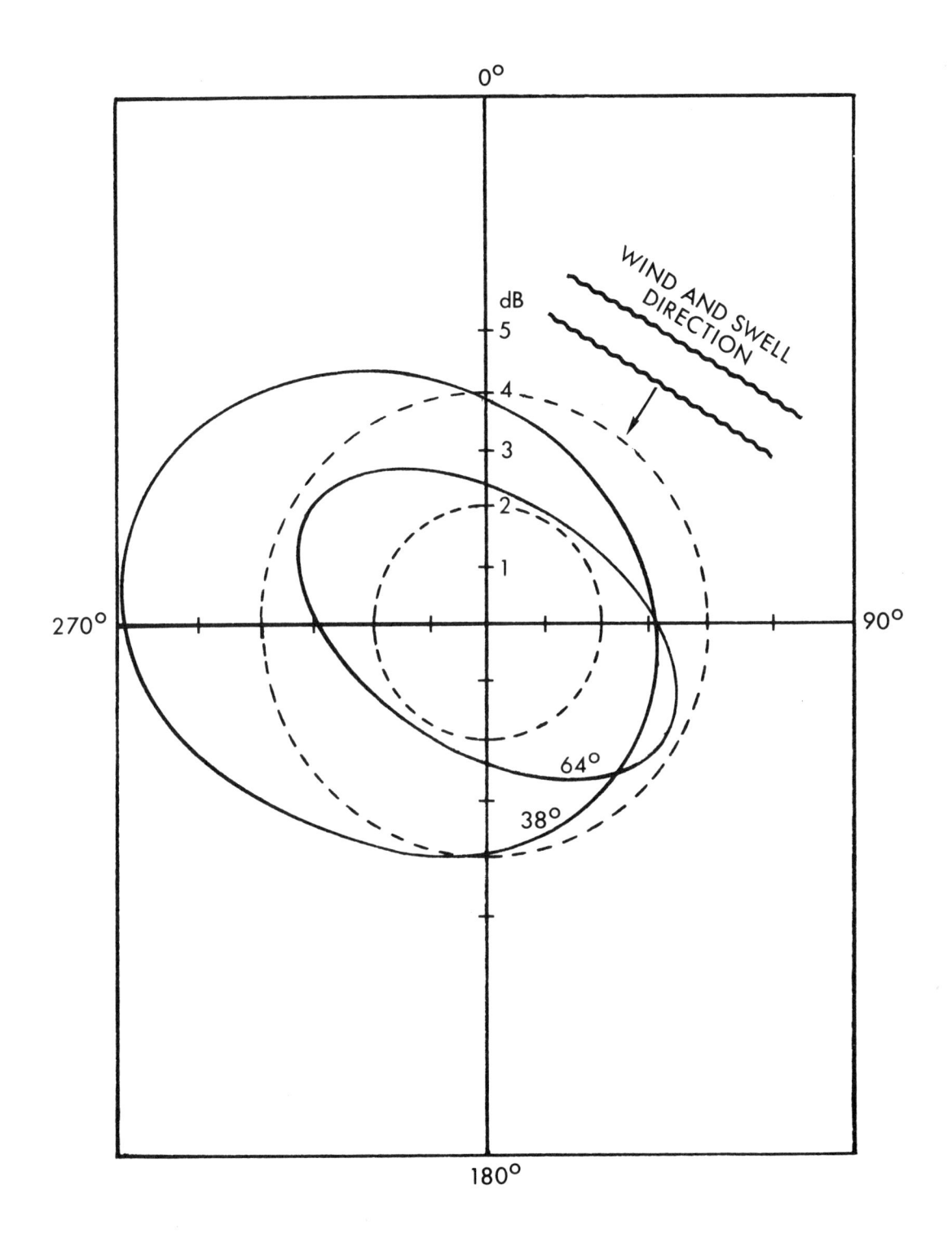

Figure 5-10. Horizontal patterns at two vertical angles. The patterns are elongated parallel to the wind and swell. The frequency band was 750 to 1500 Hz. Reference 122.

5-17

At lower frequencies, where ship noise is dominant, the horizontal directionality of the noise field depends on the source level, the transmission loss and the direction of all of the individual ships that contribute to the noise field. As an example, Figure 5-11 shows a hypothetical pattern of noise sources in which each dot represents the received level and direction of a single ship as seen from the measurement location. An array at this location would have a horizontal ambient noise beam pattern that would result from convolving its beam pattern function with the pattern of the noise sources. A very narrow-beam would resolve the noise sources; a broadbeam pattern would not.

Towed line arrays have been widely used for low-frequency horizontal directionality measurements. They have the advantages of the large aperture required for measurements at low frequencies and a low self-noise level as a result of towing the array at a great distance behind the towing ship. On the other hand, towed arrays, with their bi-conical beams, cannot separate vertical and horizontal directionalities, and, in addition, suffer from the left-to-right ambiguity common to all horizontal line arrays. Because of these difficulties, the extraction of the horizontal directionality of the noise field from beam-output measurements is both difficult and uncertain. However, Wagstaff (133) has described an iterative technique wherein an estimate of the horizontal directionality is convolved with the calculated beam pattern and compared with the measured beam output data. Wilson (134) has developed a Fourier analysis method for estimating the symmetric part of the horizontal noise field from array measurements, wherein an estimate of the noise field is made and expressed as a Fourier series having coefficients determined by the measurements. Wagstaff (135) has extended the theory from two to three spatial dimensions.

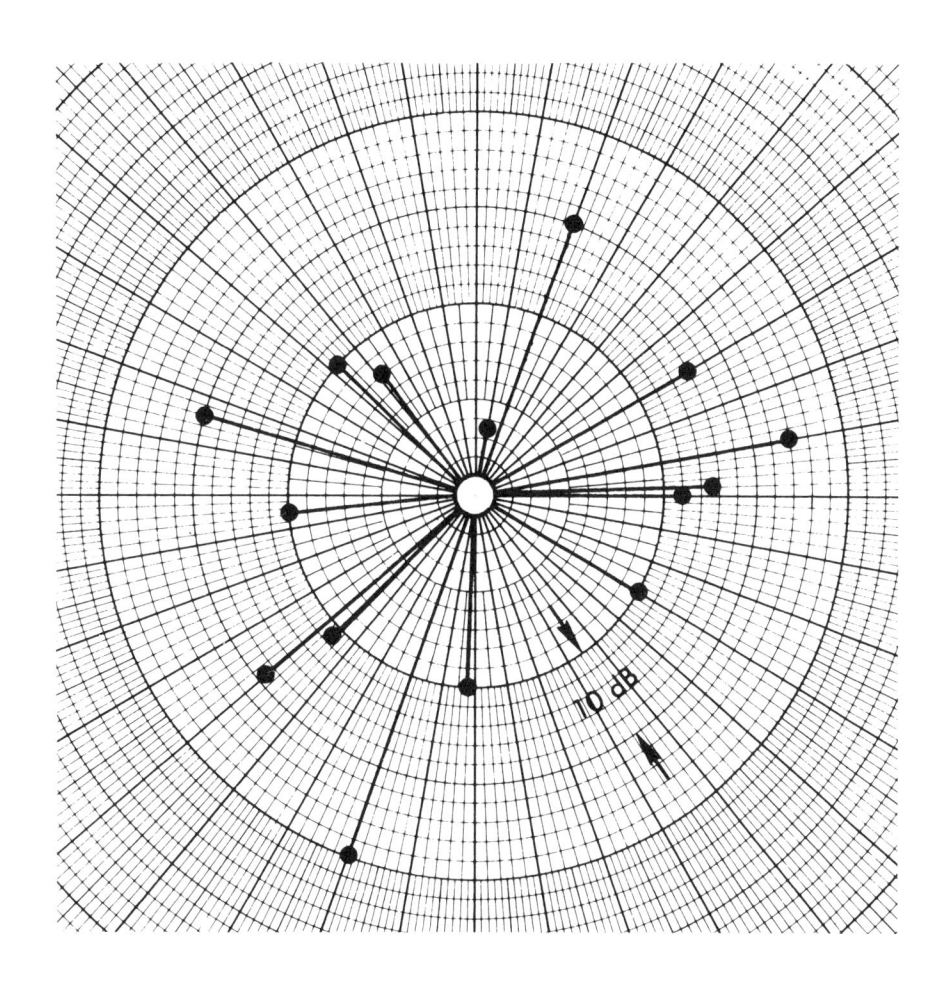

Figure 5-11. A hypothetical example of the directionality of shipping noise as it would be observed with an array of infinite resolution. The noise pattern with an actual array would be obtained by convolving the dots with the beam pattern of the array.

These theoretical difficulties can be avoided to some extent by towing an array in different directions and using the unambiguous fore and aft end-fire beams with their narrow conical patterns to obtain the desired directional data. This was done by Wagstaff and Aitkenhead (136) who towed an array in five directions 72° apart at several locations in the northeast Pacific Ocean. Figure 5-12 shows the directional pattern measured at the location marked on the chart. The lobe to the northwest was doubtless caused by ship traffic south of the Aleutian Islands passing along the great circle track to and from the Far East. This sound would suffer a low transmission loss because of the shallow depth of the axis of the Deep Sound Channel in this northerly area; the lobe to the northeast can perhaps be attributed to down-slope conversion of the sound of ships traveling on the continental slope off the west coast of the United States.

In another field trial where a line array was towed in a 6- to 12-sided polygon, reported by Wilson (134), more noise at 10 Hz was observed under high wind conditions in the near-broadside beams than in beams at other angles. This occurred for all towing directions. In calm weather, and at a frequency of 25 Hz, no high levels were observed. One possible explanation that was advanced for this effect is that wind noise at 10 Hz is spatially correlated over large distances, and that the high levels near broadside were caused by spatially correlated noise originating from the sea surface directly above the array.

A review of the problems associated with noise directionality and towed arrays may be found in a report by O'Connor and others (137).

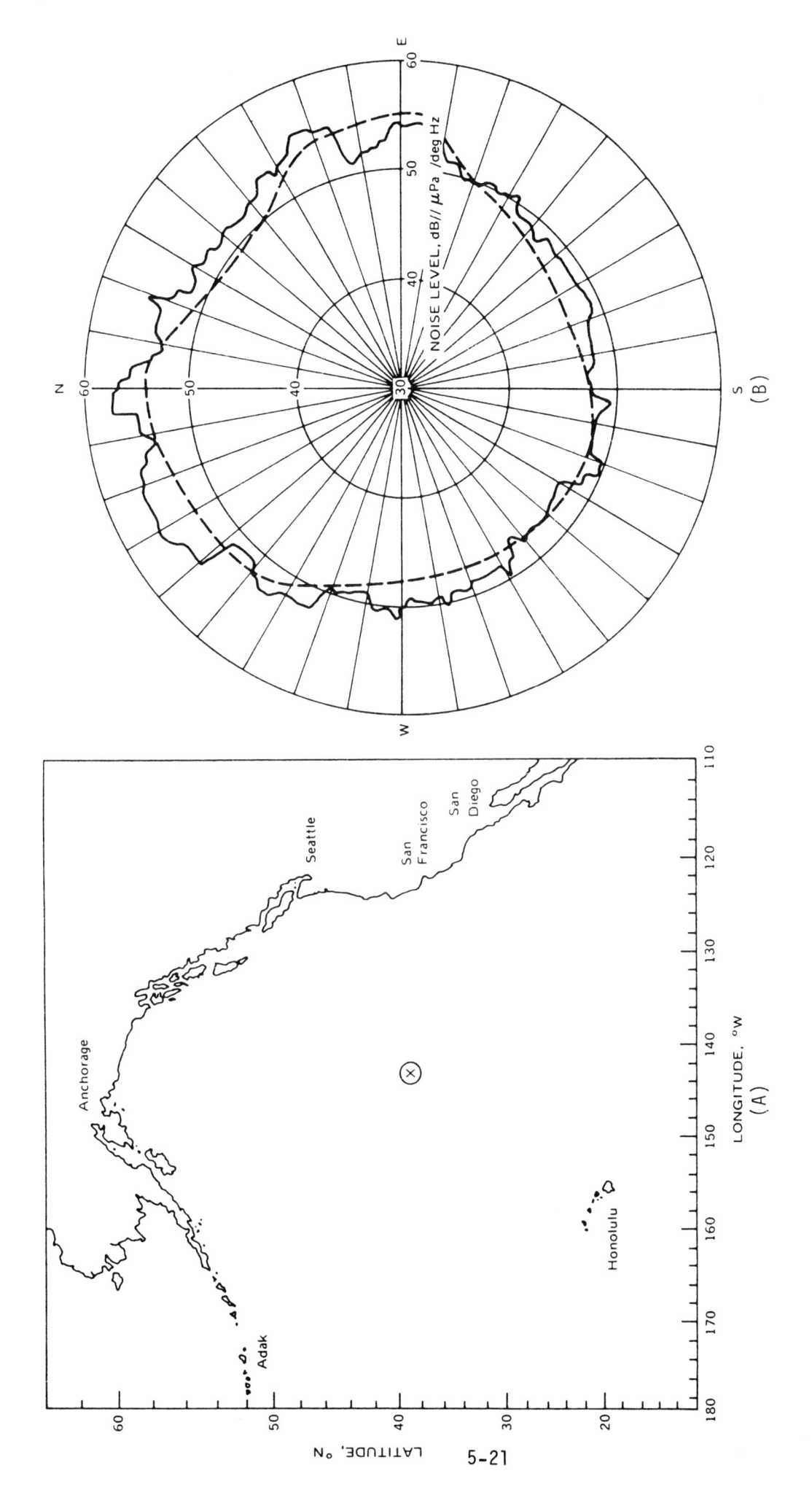

Figure 5-12. Horizontal directionality of ambient noise. (A) Site location. (B) Directional patterns, as obtained with the end-fire beam of a line array (dashed) and with broadside beams (solid). Reference 136.

5-21

5.6 __Summary__

The directionality of ambient noise in the vertical has been measured a surprisingly large number of times, all with concordant results, starting in 1961 with the work of Becken. The measurements show that at frequencies where sea-surface noise occurs, the directional patterns are oriented upward, with more noise per unit solid angle coming from the vertical than from the horizontal; at the lower frequencies of ship noise, the reverse is the case, with more noise coming from the horizontal than from the vertical. At intermediate frequencies, where there is an admixture of the two types of noise, the distribution is sensibly isotropic. The vertical directionality of ship noise varies with depth in deep water, being broader on the channel axis than away from it, with a notch or null in the directional pattern if resolution permits, due to the absence of near-horizontal rays from surface sources near the channel axis in the absence of down-slope conversion. In the horizontal, surface noise (according to a single observation) is stronger at right angles to the directions to and from which the wind and/or swell are coming. On the other hand, the horizontal pattern of ship noise depends on the distribution of ships around the measurement location and the resolution of the measurement array, there being more noise--as one would expect--coming from the directions in which there are more ships.

6. COHERENCE OF AMBIENT NOISE

6.1 Introduction

By the "coherence" of noise in the sea we mean the degree to which the noise pressures are the same at two points in the sea located a given distance and direction apart. Pressure sensors placed at the two points will have identical outputs if the noise is perfectly <u>coherent</u>; if the two sensor outputs as functions of time are totally dissimilar, the noise is said to be <u>incoherent</u>.

The degree of coherence is expressed quantitatively by the same coefficient that is used in statistics to express the degree of dependence between two variables. For two time functions $v_1(t)$ and $v_2(t)$ having zero means, this quantity is the <u>correlation coefficient</u>, or <u>normalized covariance</u>, defined by

$$\rho_{12} = \frac{\overline{v_1(t) \cdot v_2(t)}}{\left[\overline{v_1^2(t)} \cdot \overline{v_2^2(t)}\right]^{1/2}}$$

where the bars indicate time averages. For finite v's, ρ lies between +1 and -1.

While at first sight coherence may seem to be a somewhat esoteric and arcane subject, it has direct practical application to the design of hydrophone arrays. For example, it can be shown (117, p. 35) that for uniform signal and noise the gain of an unshaded array, defined as the increase of the signal-to-noise ratio - measured in dB - at its output terminals relative to that at one hydrophone of the array, is equal to

$$10 \log \frac{\sum_i \sum_j (\rho_s)_{ij}}{\sum_i \sum_j (\rho_n)_{ij}}$$

where $(\rho_s)_{ij}$ is the correlation coefficient of the signal, and $(\rho_n)_{ij}$ is the correlation coefficient of the noise, between the ith and j^{th} hydrophones. Thus, if the coefficient ρ_n is positive and the noise is therefore partly coherent between array elements, the gain of an array will be less than it would be for incoherent noise, for which $\rho_n = 0$ for all i,j except i=j.

Basic information about ambient noise directionality, discussed in the last section, can be obtained directly from the cross-correlation matrix of the outputs of a hydrophone array. This subject--the relation between directionality and coherence--is treated in a paper by Yen (138).

An example of an array gain calculation using real-world data obtained with the Trident Vertical Array has been presented by Cron and Shaffer (139). At a high wind speed (41 to 63 kn), where the noise is free of contamination by the horizontally arriving noise of distant shipping, an array gain of 15 dB can be had for an array of 8 elements spaced one-half wavelength apart, as compared with a gain of 9 dB (10 log 8) that would be obtained with isotropic noise and a perfectly coherent signal.

6.2 Theory

The simplest kind of noise to deal with theoretically is _isotropic_ noise, for which the correlation coefficient can be obtained by a simple integration. For the case of a single frequency uni-directional plane wave incident at an angle θ to the normal between two hydrophones spaced a distance d apart, the correlation coefficient is easily found to be $\rho_\tau = \overline{\cos \omega\tau}$, where $\omega = 2\pi$ times the frequency and $\tau = \frac{d}{c} \sin \theta$, the travel time of a plane wave between the two hydrophones. If this equation is integrated over θ and normalized, the result is, for isotropic noise at a single frequency,

$$\rho = \frac{\sin kd}{kd}, \text{ where } k = \frac{2\pi}{\lambda}$$

This function falls to zero, and noise becomes uncorrelated, at multiples of a half wavelength. Because of this, many arrays use, as an approximation, a spacing of $\lambda/2$ between elements at the mid-band frequency. Complicated expressions are found (140) when this simple equation is integrated over a band of frequencies and when an electrical time delay--corresponding to the steering of a beam--is inserted. A table listing these expressions is given in Figure 6-1.

However, ambient noise is seldom isotropic, except when it is a happy combination of ship and surface noise. At high frequencies, a more appropriate theory that applies for sea surface noise considers the case of a dense distribution of random noise sources lying on a plane surface. The theory for this case was first worked out by Cron and Sherman (141) for pressure, or amplitude, radiation patterns of the surface sources expressed by the relationship $g(\theta) = \cos^m\theta$, where $g(\theta)$ is the pressure amplitude radiated by a surface element at an angle θ to the vertical, normalized so that $g(0) = 1$. For a dipole radiator, $m = 1$. The results of these calculations for values of $m = 0, 1, 2$ and 3, as well as the corresponding curve for isotropic noise, are given in Figure 6-2. On close inspection it will be seen that as m increases and the surface sources become more directional, the first zero of ρ occurs at an increasingly greater spacing and the amplitude of the secondary peaks decreases. More recently, an alternate approach using spatial harmonics has been formulated by Cox (142), and good agreement with experimental data over a wide frequency range was found.

The addition of an electrical time delay, required for the formation and steering of a directional beam, greatly complicates the theory, which appears to have been first worked out by Jacobson (140).

	Unidirectional signal ρ_s	Isotropic noise ρ_n
Single frequency, zero time delay	$\cos \omega \tau_w$	$\dfrac{\sin(\omega\, d/c)}{\omega\, d/c}$
Single frequency, time delay	$\cos \omega(\tau_w + \tau_e)$	$\dfrac{\sin(\omega\, d/c)}{\omega\, d/c} \cos \omega \tau_e$
Flat bandwidth, zero time delay	$\dfrac{\sin[\tfrac{1}{2}(\omega_2 - \omega_1)\tau_w]}{\tfrac{1}{2}(\omega_2 - \omega_1)\tau_w} \cos \dfrac{(\omega_2 + \omega_1)}{2}\tau_w$	$\dfrac{1}{(\omega_2 - \omega_1)d/c}\left[\operatorname{Si}\!\left(\omega_2 \dfrac{d}{c}\right) - \operatorname{Si}\!\left(\omega_1 \dfrac{d}{c}\right)\right]$
Flat bandwidth, time delay	$\dfrac{\sin[\tfrac{1}{2}(\omega_2 - \omega_1)(\tau_w + \tau_e)]}{\tfrac{1}{2}(\omega_2 - \omega_1)(\tau_w + \tau_e)} \cos \dfrac{(\omega_2 + \omega_1)}{2}(\tau_w + \tau_e)$	$\dfrac{1}{2(\omega_2 - \omega_1)d/c}\Big[\operatorname{Si}\!\left(\omega_2(\tfrac{d}{c} + \tau_e)\right) - \operatorname{Si}\!\left(\omega_1(\tfrac{d}{c} + \tau_e)\right)$ $+ \operatorname{Si}\!\left(\omega_2(\tfrac{d}{c} - \tau_e)\right) - \operatorname{Si}\!\left(\omega_1(\tfrac{d}{c} - \tau_e)\right)\Big]$

Symbols for table:
$\omega = 2\pi$ (frequency)
τ_w = travel time of signal between array elements
 $= (d/c) \cos \theta$
d = separation of array elements
θ = angle to line joining the two elements

c = velocity of sound
τ_e = electrical (steering) delay
$\omega_2 = 2\pi$ (upper frequency of band)
$\omega_1 = 2\pi$ (lower frequency of band)

$$\operatorname{Si}(x) = \int_0^x \dfrac{\sin u}{u}\, du$$

Figure 6-1. Expressions for the coherence of unidirectional and isotropic noise for a single frequency and for a band of frequencies, with and without an electrical time delay. Reference 117, Table 3.1.

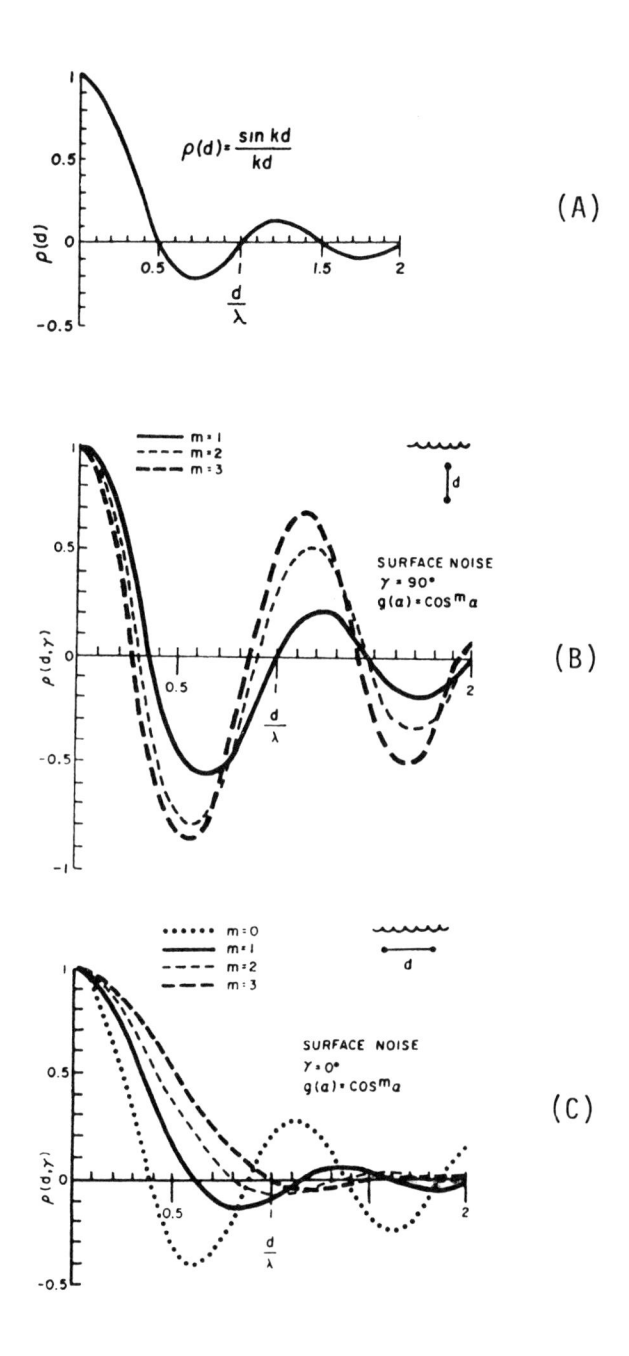

Figure 6-2. Coherence functions vs. normalized spacing. (A) Isotropic noise. (B) Surface noise with vertical spacing. (C) Surface noise with horizontal spacing. The parameter m applies to the amplitude radiation pattern of the surface sources. Reference 141.

The theoretical results can be best expressed by contour maps of ρ on coordinates of normalized spacing d/λ vertically and normalized time delay $\frac{\tau}{d/c}$ horizontally. Such plots were computed by Liggett and Jacobson (143), and in more detail by Edie (144). An example of such contours is given in Figure 6-3 for a value of m = 1, along with the results of an experimental investigation (145), yielding time-delay correlograms for a number of vertical hydrophone separations. A somewhat similar theoretical plot on similar coordinates has been published by Lytle and Moose (146) for a frequency band f_2/f_1 = 1.7.

Instead of assuming a directional model for the surface radiators, a theory has been developed by Anderson and Tittle (126) on the basis of an assumed known covariance of the shape of sea surface--from which the covariance at any two points in the sea below can be obtained. The results are in general agreement with the results of the first model. Another theoretical treatment (147) considers the case of two vertically separated hydrophones, one close to the sea surface, the other at a depth, with the finding that the vertical correlation of surface noise depends on the spacing of the hydrophones just as it does at great depths, even when one or both hydrophones are close to the surface and the ratio of depths is large compared to unity. Another effect discussed in the theoretical literature by Liggett and Jacobson (148) is the effect on the correlation of attenuation and spherical spreading. These same authors have also investigated (149) the space-time covariance of surface noise by means of still another model that assumes that the covariance of pressure is specified for all pairs of points on a plane surface and that the acoustic pressure satisfies the wave equation throughout the medium. The results agree with those found by earlier investigators who assumed a random distribution of directional surface sources.

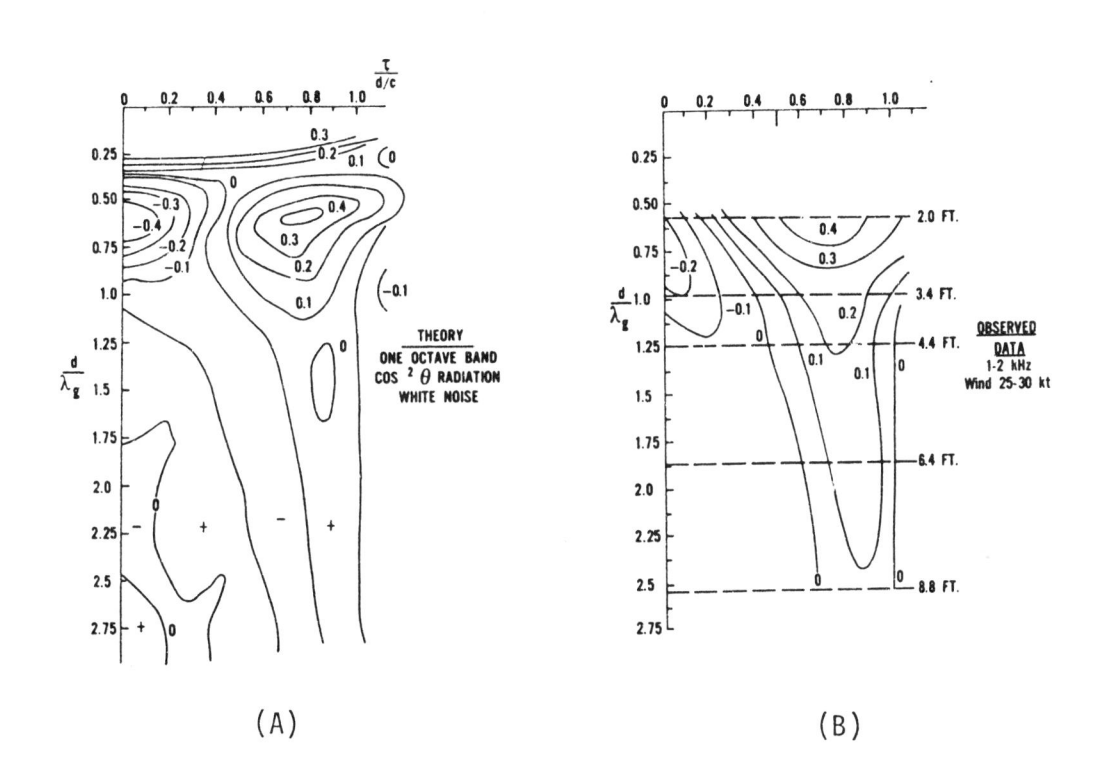

Figure 6-3. Correlogram contour plots showing the correlation coefficient as a function of normalized spacing and time delay. (A) Theoretical plot for a one octave frequency band and for dipole radiation. (B) Contours based on measured data for the 1 to 2 kHz band for values of d from 2.0 to 8.8 ft. Reference 145.

In all of the foregoing, the bottom has been neglected. Several papers discuss the theory of noise coherence in shallow water; among these may be mentioned papers by Buckingham (150) (151) (152), and Kuperman and Ingenito (153), and an interesting paper on the gain of arrays in shallow water by Lloyd and Daintith (154).

For application to the design of line arrays for the detection of ships travelling in shipping lanes. Sloboda and Manness (223) have worked out theoretically the horizontal coherence at various angles from $\theta_o=0$ (parallel to the lane) to $\theta_o=\pi/2$ (perpendicular to the lane) of the noise from a distant infinitely long line source. A comparison of theory and observations in the northeast Pacific Ocean was given for a location where the dominant source of noise is distant ship traffic travelling along well-defined transoceanic shipping lanes. The theory shows that the coherence function as its first minimum at $d/\lambda=0.3$ when $\theta_o=0$, moving out to 0.7 for $\theta_o=\pi/2$.

6.3 Field Observations

A vertical array of 40 hydrophones moored near the bottom in 14,400 ft of water has been the tool for a number of studies of noise coherence. This array, called the Trident Vertical Array (TVA), described earlier in Section 5.2, is shown in Figure 6-4. An example of the results obtained in an experiment by Urick (145) with this array has been seen in Figure 6-3b, where there is good agreement between observations and theory (Figure 6-3a) in the band 1 to 2 kHz for sea state 4 with m = 1 (that is, the surface dipole model). The peak value of the coefficient occurs at a time delay corresponding to the hydrophone spacing for noise coming from above, as it should. This finding, together with the fact that at 500 Hz the correlation coefficients at

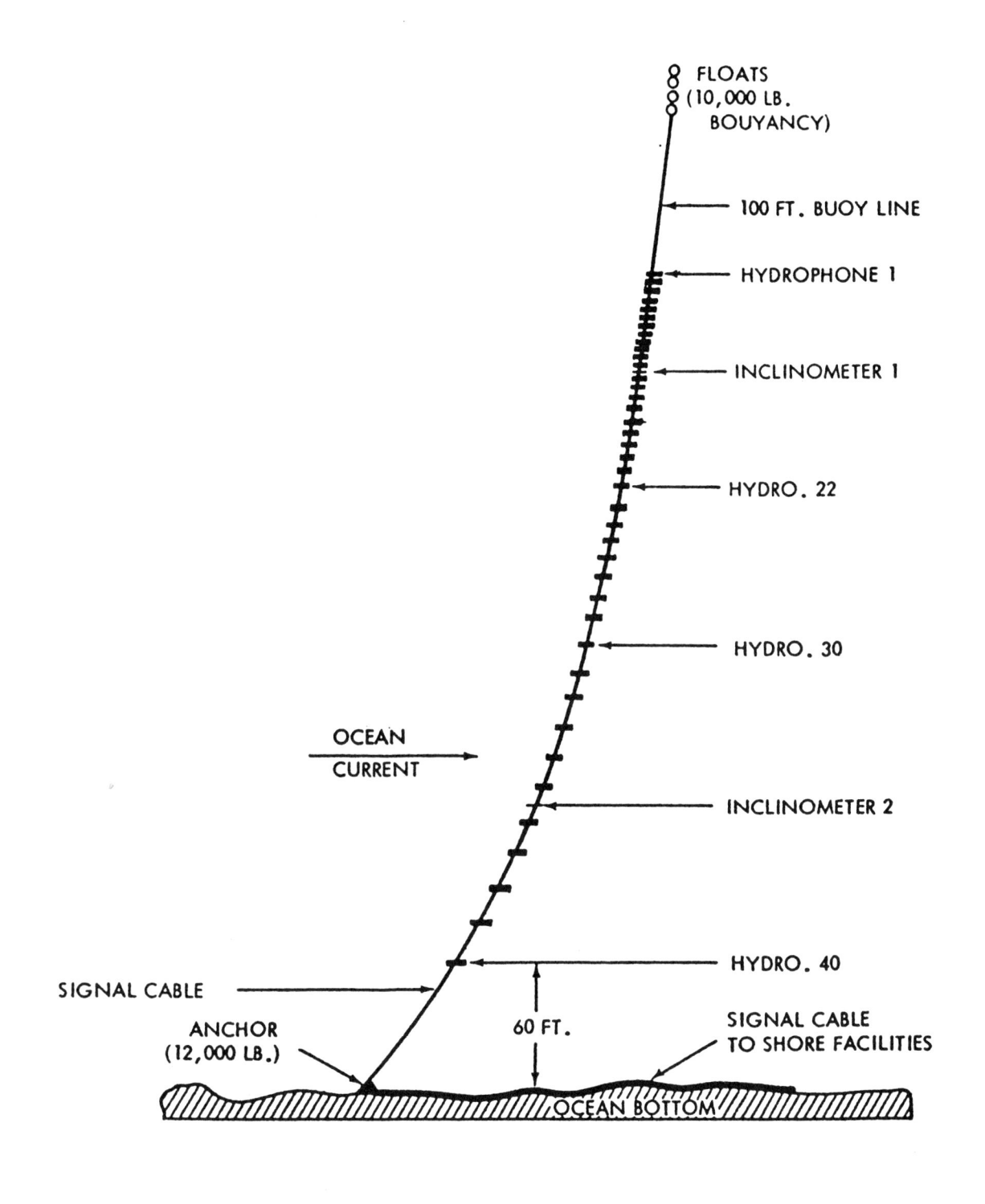

Figure 6-4. The Trident Vertical Array at Bermuda, with which a number of coherence measurements by different investigators have been obtained. Reference 218.

different spacings can be normalized by dividing by the wavelengths, was confirmed in several papers by Arase and Arase (155) (156) (157).

On the other hand, at a lower frequency and at a lower sea state (200 to 400 Hz, sea state 3), the agreement between theory and observations was found to be poor (158), doubtless because of contamination by the noise of distant shipping, which, arriving nearly horizontally, cannot produce negative values of the correlation coefficient between pairs of a vertical array, as required by the simple model.

The Bermuda array has given repeated confirmation that the sea surface acts, as far as coherence is concerned, like a distribution of random dipoles when both sea state and frequency are high (157) (145). At lower frequencies in the region of shipping noise, there seems to be only a small amount of data obtained with this array. One example of low-frequency coherence data obtained with the array is given in Figure 6-5, where the correlation coefficient is seen to be only poorly normalized in terms of d/λ_m, the ratio of vertical spacing to the mid-band wavelength. Nevertheless, the coherence is much greater at low frequencies than it is at high frequencies for a spacing of the same number of wavelengths. In short, shipping noise, arriving nearly horizontally and therefore highly collimated, is much better correlated between vertical pairs of hydrophones than is surface noise arriving from above.

Time-delay correlograms obtained at Bermuda show graphically the profound distinctions between the two types of noise. In Figure 6-6 the left-hand vertical string of correlograms were obtained for a wind speed of 4 knots, the right-hand series for a wind speed of 30 knots. The shipping noise correlograms at the left show a correlogram peak that remains high and zero-centered with increasing separation. On the other hand, the wind-noise

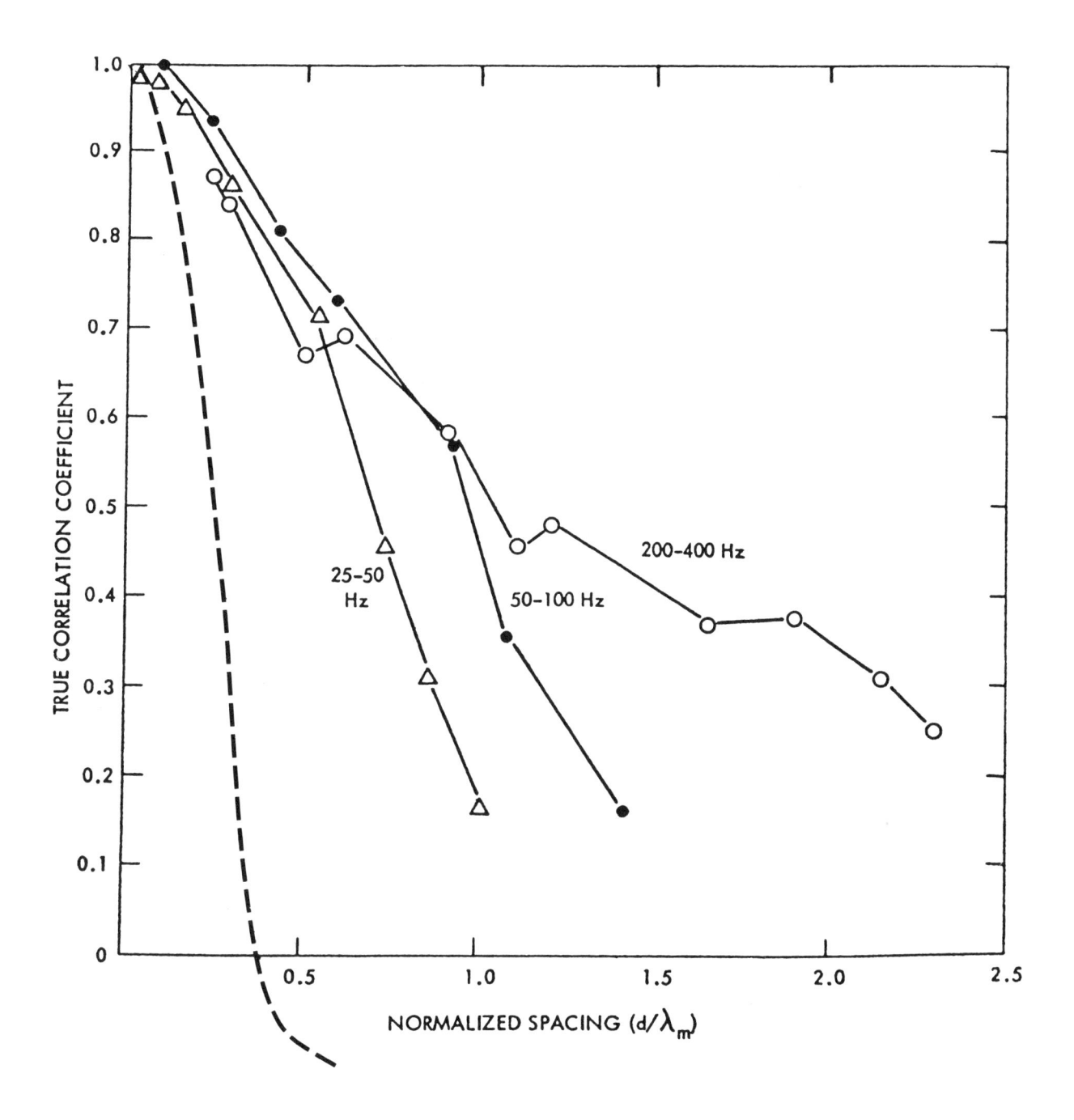

Figure 6-5. Octave band low frequency coherence of ambient noise as found with the Trident Vertical Array, normalized to the mid-band wavelength λm. The dashed curve is the theoretical curve for surface noise for $m = 1$ (Figure 6-2B); the better coherence of low-frequency noise is evident. Reference 218.

correlograms at the right show a peak that is weaker and moves out in time-delay as the separation increases. That is, the correlograms are of two types, in strict agreement with our two-component model of deep-sea noise.

If we call these two types of correlograms Type I and Type II, corresponding to shipping noise and wind noise, and plot additional correlogram data on coordinates of wind speed and frequency, we obtain Figure 6-7 where the correlograms for each combination are labeled I or II. Intermediate correlograms were obtained at points labeled I + II. The cross-hatched band shows the boundary on these coordinates between the regions of dominance of the two types of noise at the Bermuda location.

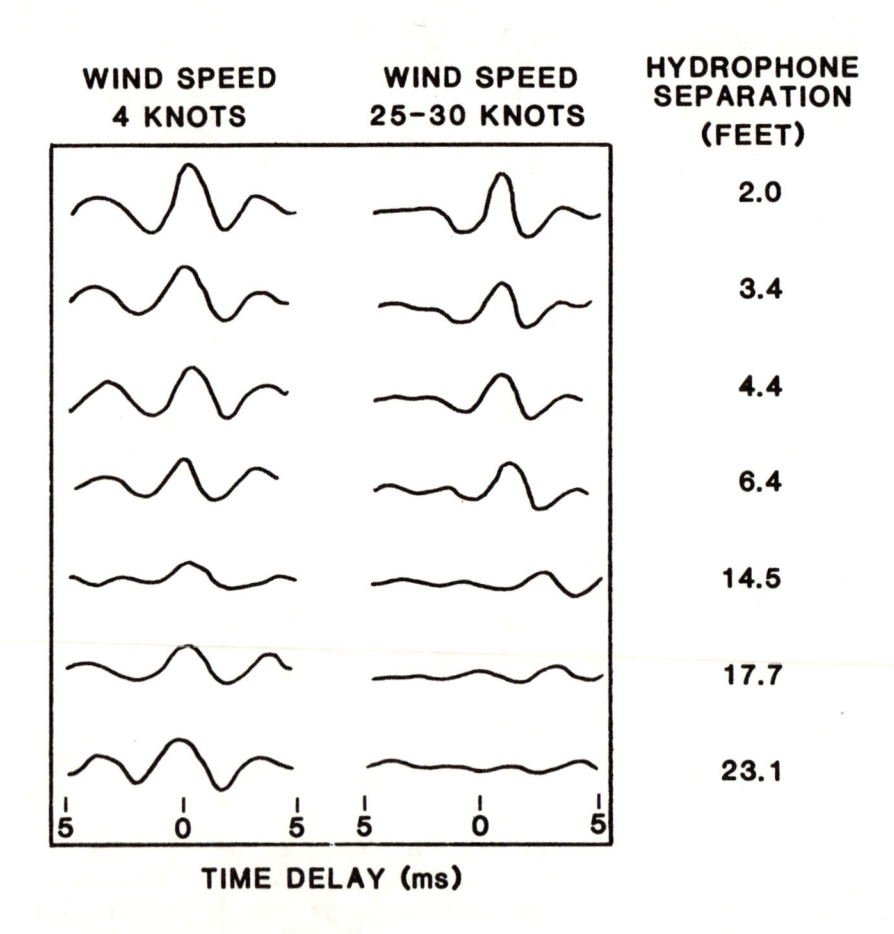

Figure 6-6. Time-delay correlograms at a high and a low wind speed in the 200-400 Hz octave band for various vertical hydrophone separations. Reference 117, Figure 7-26.

6-12

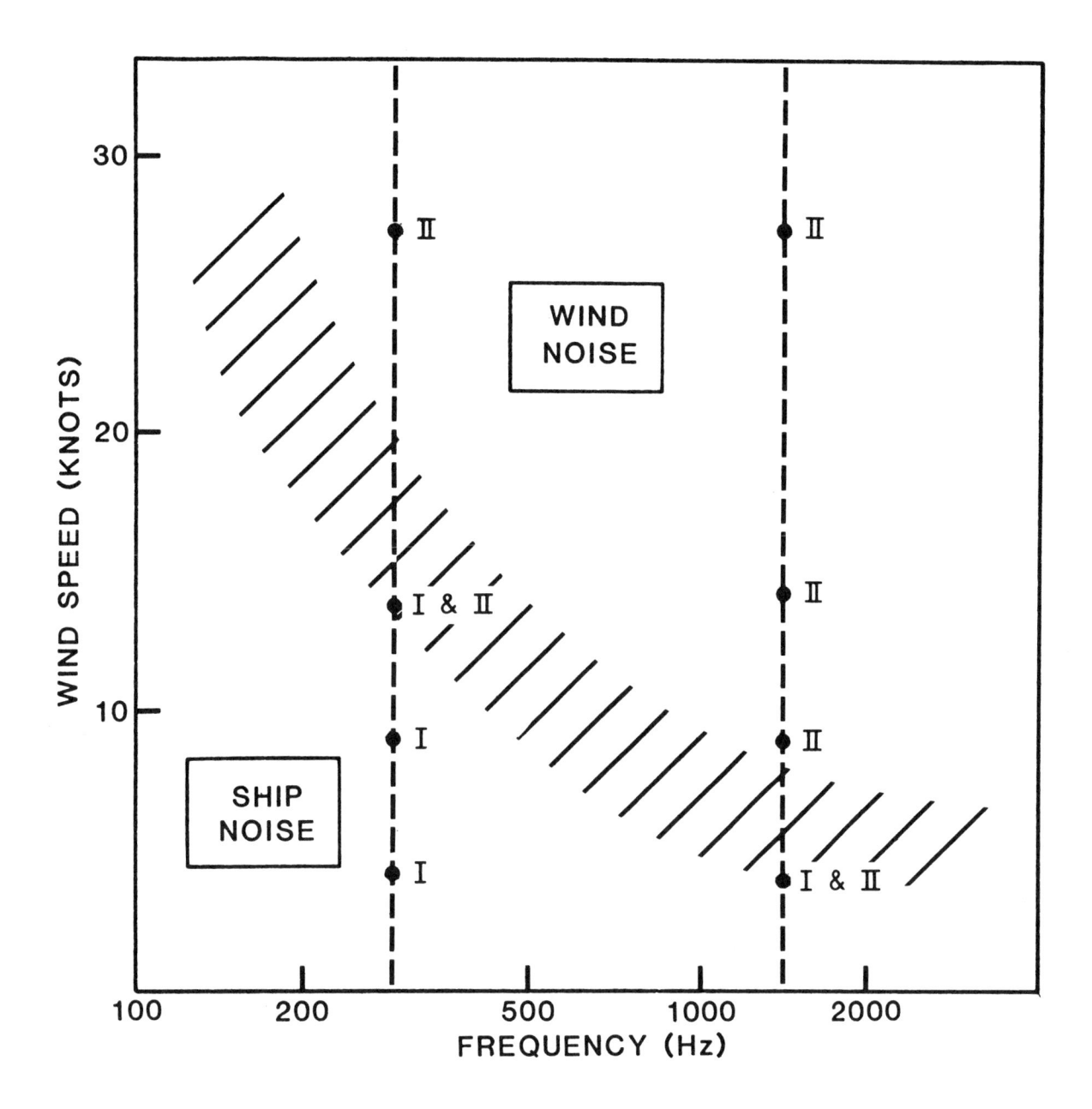

Figure 6-7. Regions of dominance of ship and wind noise as deduced from the types of correlograms obtained. Reference 145.

The noise in a surface duct also appears to have a two-component origin, one component originating locally and arriving from near-vertical directions, the other originating at a distance and arriving horizontally. Correlograms obtained by Urick (116) in a 120-foot surface duct are shown in Figure 6-8 for the frequency band 1 to 2 kHz.

Other arrays at other locations have been used for noise coherence measurements. Urick (159) investigated the effect of depth on coherence at St. Croix; no variation of noise coherence with depth was observed. In another experiment, Linnett and Thompson (160) used a pair of vertically separated hydrophones 13.2 ft apart located close to the bottom in 360 ft of water; the results were consistent with deep water theory for the cosine source model with m = 1 or 1-1/2 at wind speeds of 10 to 25 knots, and with m = 1/2 for wind speeds from 2 to 5 knots.

6.4 <u>Summary</u>

The coherence of ambient noise between vertically separated hydrophones has been repeatedly found to agree with the theory of a random dense surface distribution of noise sources. For mathematical simplicity, a theoretical amplitude radiation pattern, $g(\theta) = \cos^m \theta$, has generally been assumed, and the observations tend to agree that m = 1, corresponding to a surface distribution of dipoles. Other values of m have fitted some data, although the possible contamination by shipping and other noises in these cases cannot be eliminated. Only a small amount of data exists for shipping noise correlated between vertical hydrophones, and nothing at all has been done on the coherence of ambient noise in deep water between horizontally spaced hydrophones. The importance of the subject lies in its relationship to the design of arrays to discriminate against ambient noise, where it forms the basic input data for computer programs using trial array designs.

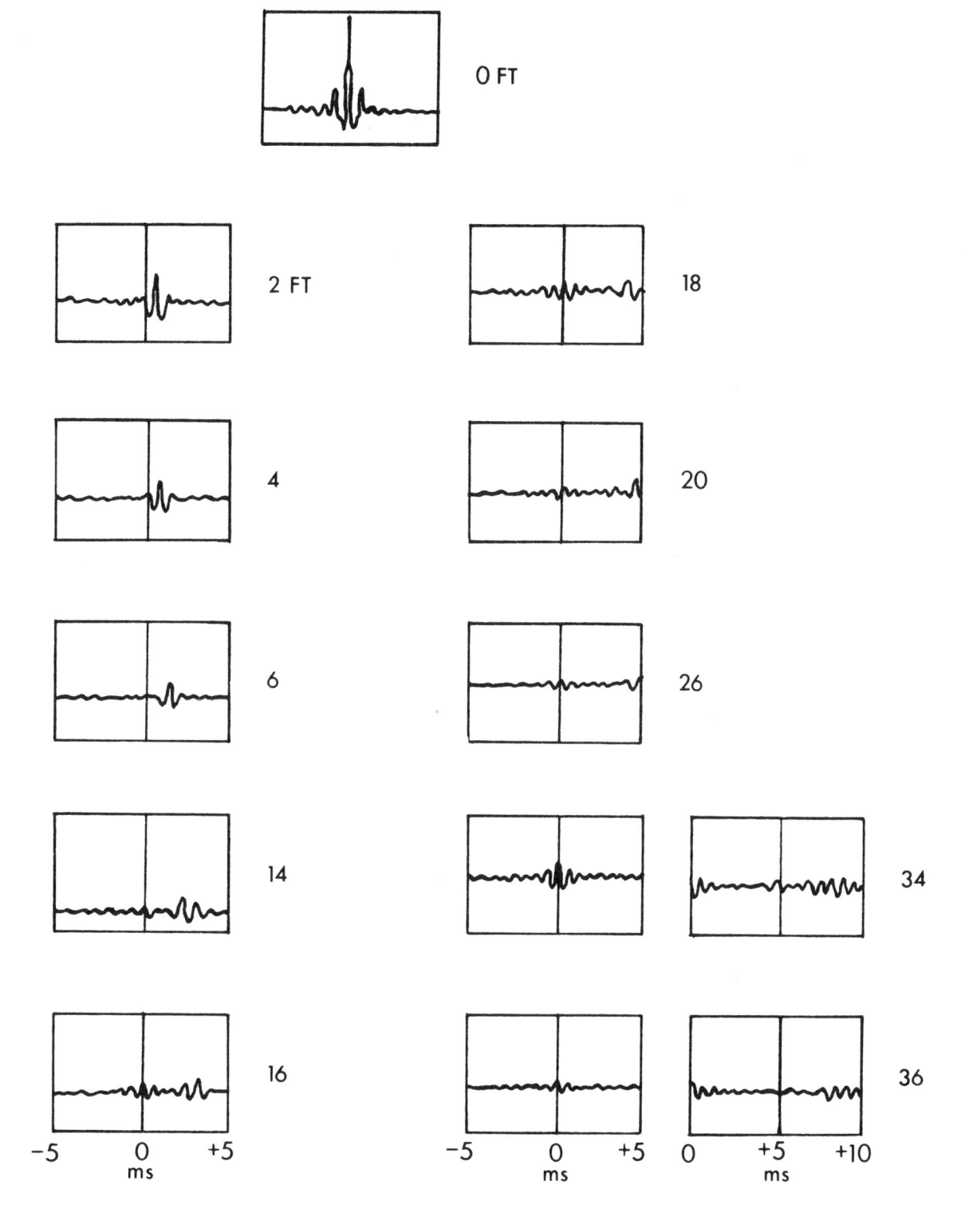

Figure 6-8. Correlograms at different vertical separations in a surface duct, showing two types of noise. The frequency band is 1 to 2 kHz. The top correlogram (zero separation) is an autocorrelogram that calibrates the two scales of the other correlograms.
Reference 116.

7. INTERMITTENT SOURCES OF NOISE

7.1 Introduction

By intermittent sources we mean those noises, while at times occasional and irregular in occurrence, that are sometimes persistent and regular enough to be considered part of the ambient background. Such intermittent sources of noise may be divided into Biologic Noises caused by marine animals and Non-biologic Noises, such as the underwater noise made by rain from above and earthquakes from below.

7.2 Biologic Noises

When you listen with a broadband hydrophone at a biologically active location, you will likely hear a weird cacophony of noises such as whoops, grunts, whines, moans, chirps and many others. Such sounds are made by one or more species of marine animals.

The various soniferous species in the sea belong to one of the three major classes: the crustacea (or shellfish), the marine mammals (such as the whales) and certain species of true fish (such as croakers). Strangely, no kind of commercially sought-after fish, such as cod, plaice, flounder, etc., appears to make noise. Incidentally, this is a great boon to the survival of the commercial species, for fish schools could be easily found and fished for--if they did make noise--by listening with a simple directional hydrophone.

Strange to say, the fact that marine animals make noise was not a subject of scientific study until the advent of World War II. At this time, the importance of biologic noise to the operation of the wartime sonars and acoustic mines first came to be recognized. In subsequent years biologic noise has become a favorite subject in the acoustic and biological literature; a paper by Backus (161), dated 1958, contains 70 citations to the literature.

Nearly all species of fish make noise—if not in their natural environment, then when they are caught and subject to stimuli such as electric shocks.

Biologists have long speculated (161) about the reasons that animals in general, and marine animals in particular, make noise. One is <u>communication</u>; whales are said to communicate between members of a school. Another is <u>defense</u>, attempting to scare away attackers. Another is <u>location</u>, as in the echo-ranging of porpoises. Another is <u>sexual attraction</u>; most fish are noisy only within the male sex. A final reason must be that it must be <u>pleasurable</u> for a marine animal to make noise, just as in the humming of a tune by a human being.

Two books on the subject of biological noise in the sea are worthy of special mention. One is a compilation of papers presented at a conference held in 1963 (162); the other (163) is an analytic description, with many oscilloscope photographs, of the sounds made by North Atlantic fishes.

In what follows, we will restrict our discussion to three categories of soniferous marine animals: (1) <u>whales, dolphins and porpoises</u>, (2) <u>croakers or drum fish</u>, and (3) <u>snapping shrimp</u>. We will give separate attention to the once mysterious "twenty-cycle pulses"—certain unique sounds now known to be made by a certain species of whale.

7.2.1 Whales, Dolphins and Porpoises

Dolphins and porpoises are small whales that are distinguished from one another by the shape of the nose or beak (the nose of dolphins is pointed, that of porpoises is blunt). The noise made by all the whales is highly variable and depends not only on the species, but, presumably, upon the purpose or reason, if any, for the animal to make the noise. They range from low-frequency moans and screams to high frequency repetitive clicks. Although nothing definite was known about whale sounds before World War II, the literature on the subject has now grown to enormous proportions; a short paper written 20 years ago by Schevill (164) contains 63 references. Because of the variety and complexity of the sounds made by whales, only a few of the more recent papers may be mentioned.

A Russian paper (165) described the whistle-type signals made by dolphins for communication purposes. Forty-nine types of signal were identified, ranging in frequency from 2 to 60 kHz and having durations between 2 ms and 2 sec. In an American paper (166), the moans made by blue whales--among the largest living creatures--were recorded off the coast of Chile and were found to have a band level at one meter of 188 dB in the 14 to 222 Hz band. In a tank, the echolocation signals made by dolphins were found (167) to be directional, with peak energies above 100 kHz. During field observations lasting two days at a deep-water location south of Bermuda (168), repetitive pulses, with a repetition period near 110 ms and with a broad spectrum centered at 3 kHz, were recorded and attributed to sperm whales. Finally, with a new technique using calibrated sonobuoys dropped by a helicopter ahead of herds of whales and dolphins, the first quantitative measurements of the source level of the sounds of herds of these animals have been obtained (169). As an example, Figure 7-1 shows the source level (measured levels reduced to 1 m) in a 120-Hz measurement band of the whistles and clicks of the common dolphin, _Delphinus delphis_. A better idea of the complexity and variety of the sounds made by whales may be had from Figure 7-2, a sound

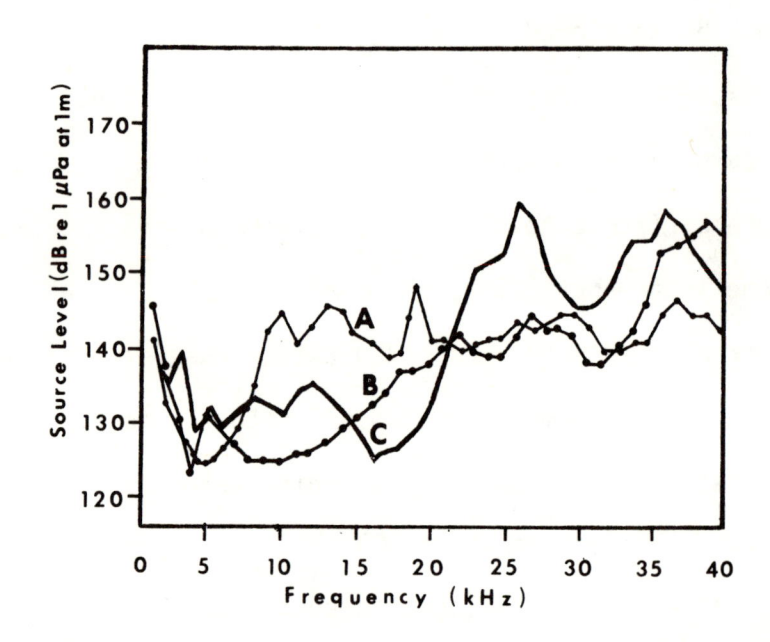

Figure 7-1. Mean levels at 1 meter, in a 120 Hz band, of the whistles and clicks from herds of the common dolphin Delphinus delphis. A - whistles, 300 animals; B - clicks, 300 animals; C - clicks, 80 animals. Reference 169.

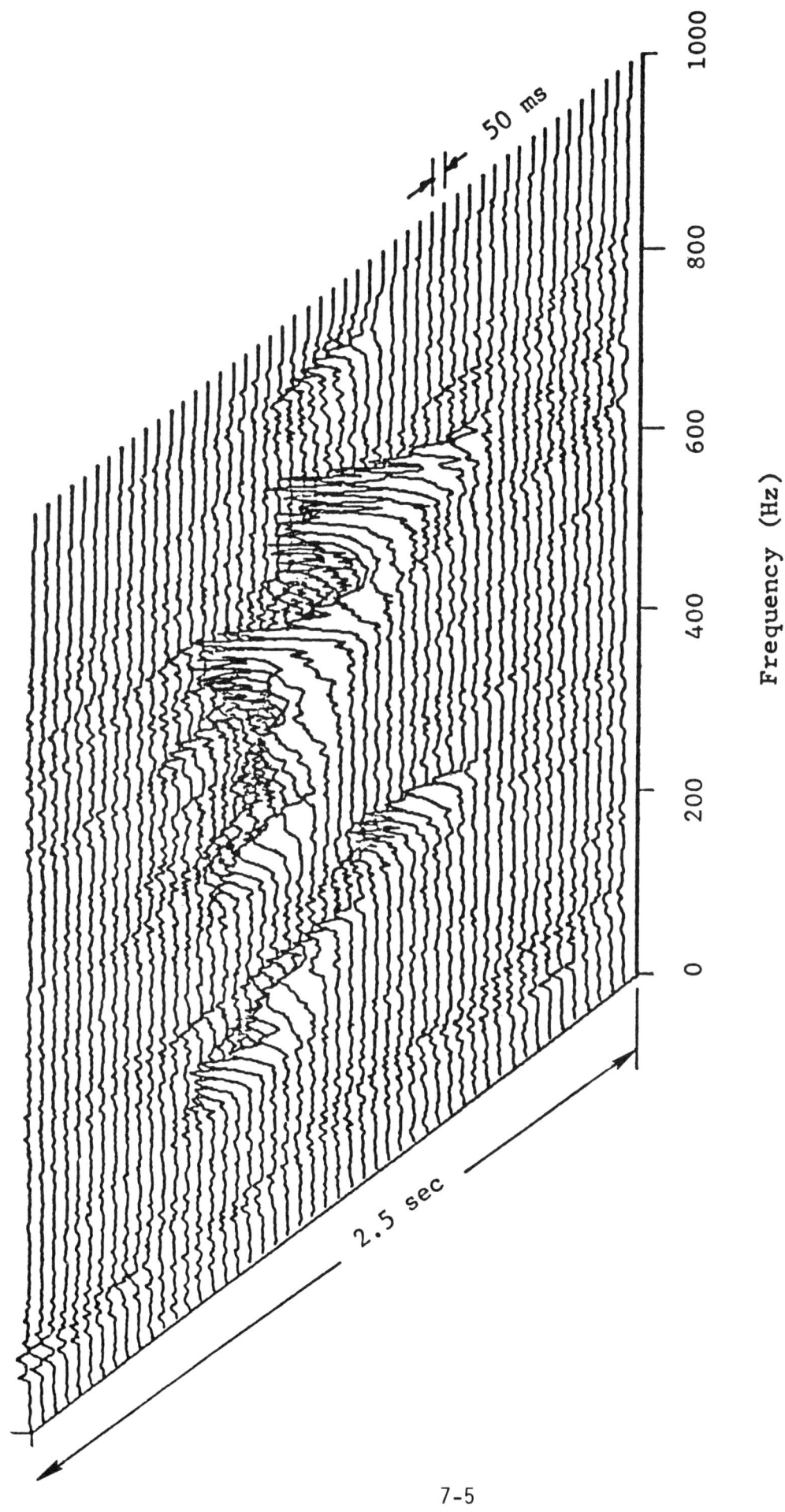

Figure 7-2. Sound spectrogram of a humpback whale. Each successive spectrum covers a time interval of 50 ms. Courtesy D.V. Holiday, Tracor, Inc., San Diego, CA.

spectrogram covering 2.5 seconds of time. Over this brief period of time, the humpback whale evidently makes a large variety of diverse sounds.

7.2.2 Croakers

A kind of biologic noise that once had great military significance is the peculiar noise made by croakers or drumfish. This noise came as a surprise during the first tests of acoustic mines during World War II in Chesapeake Bay. Strange to say, no prior observation of the underwater noise now known to be of biologic origin had appeared anywhere in the literature. The troublesome noise of croakers in Chesapeake Bay was described in one of the earliest papers on underwater sound published in the Journal of the Acoustical Society of America. The abstract of this early paper, by Loye and Proudfoot, appearing in the October 1946 issue (170), said that

> "the widespread use of underwater acoustical devices during the recent war made it necessary to obtain precise information concerning ambient noise conditions in the sea. Investigations of this subject soon led to the discovery that fish and other marine life, hitherto classed with the voiceless giraffe in noisemaking ability, have long been given credit for a virtue they by no means always practice. Certain species, most notably the croaker and the snapping shrimp, are capable of producing noise which, in air, would compare favorably with that of a moderately busy boiler factory."

The paper goes on to describe some of the interesting experiments that were carried out to establish the cause of this suprisingly intense interfering background.

Croaker noise has been likened to the sound of fast hammering on a hollow log, and is now known (171) to be made by resonant vibration of the fish's swim bladder. The noise is restricted mainly to the 350 to 1500 Hz band and exhibits a diurnal and seasonal variability. The diurnal variability has been dignified by the term "evening chorus," referring to an increase in sonic activity beginning near sunset and persisting through the early evening hours.

Croaker-like sounds have been heard in many parts of the world. In the Timor Sea north of Australia, a variety of biological sounds in the audio frequency band 100 Hz to 2 kHz have been described, apparently caused by some unidentified species of fish (172). Figure 7-3 shows spectra of the noise observed at this location along with the croaker noise in the months of May and July as measured earlier by Loye and Proudfoot (170) in Chesapeake Bay. At three other locations near Australia, evening choruses were also observed (173), though with peculiar and different characteristics. One possible source was said to be sea urchins, of common occurrence in the shallow waters of the region. These choruses had most of their energy between 400 and 4000 Hz, with levels up to 30 dB above the usual background noise. Finally, near a tower in 60 ft of water off San Diego, California (174), an evening chorus in the 100 to 1000 Hz band appeared at sunset and, to a lesser extent, at sunrise, raising the level of 5 to 20 dB above normal. At this site another sound called the "cycling sound" was observed at night in the late spring and summer seasons. It was believed to be caused by carpenter fish, members of the croaker family.

An interesting occurrence illustrating the variability of biologic noise--though not made by dolphins--was an increase in noise level of 50 dB observed (175) in the Salton Sea, California--an inland salt-water lake. This tremendous increase in broadband noise, with a peak near 1 kHz, was caused by a single species of fish--the orangemouth corvina--and occurred during a four-month period coinciding with the breeding season.

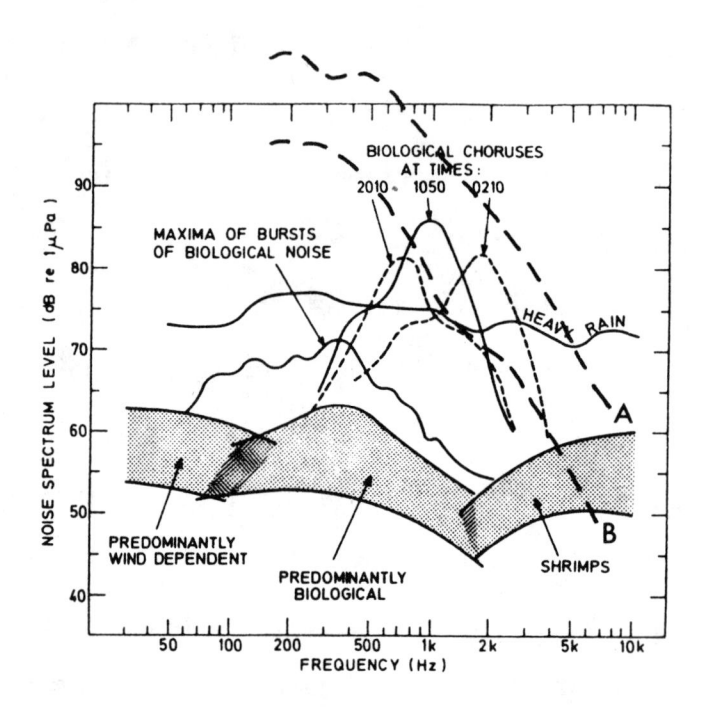

Figure 7-3. Summary of the components of ambient noise as observed in the Timor Sea north of Australia. Reference 172. The dashed curves show the spectra of croaker noise measured in Chesapeake Bay in 1942 during May (A) and July (B). Reference 170.

In a remote isolated coastal lagoon off central Baja California, low-pitched, undulating grunting sounds were heard (176) that were attributed to the Pacific gray whale--a frequent visitor to the area during the calving season.

7.2.3 Snapping Shrimp

Early in World War II, it was observed that submarine echoes in the 24-kHz echo-ranging sonars of the time died away into a high background of noise. This noise was an annoying and serious limitation to sonar ranges in the warm, shallow areas of the Pacific where many sonar operations had to be carried out. This high-frequency noise, soon found to be caused by snapping shrimp (not to be confused with the edible shrimp), completely dominated the ambient noise background in many areas.

Of the many genera of "snapping shrimp," only two species are capable of vigorous snapping. Figure 7-4a is an oscillogram of a single snap made by the single large claw of the common bay shrimp, Crangon californiensis. The spectrum of the noise made by many such snaps has a broad peak in the region 2 to 15 kHz (177).

Shrimp noise is likely to be found in waters less than about 55 m deep and warmer than about 11° C on bottoms composed of shell, coral, rock or other hard sheltering materials. The noise occurs extensively in tropical and sub-tropical coastal waters throughout the world. It has a diurnal variability, being greater at night by 3 to 6 dB (177).

In one study (48), shrimp noise was observed in a coastal lagoon 6 fathoms deep off Baja California along with a variety of other biological noises, as well as in the ports of San Diego and Long Beach in California, and Balboa and Cristobal in the Canal Zone. Also, shrimp noise has been reported and measured (178) at a number of shallow locations off Miami, Florida.

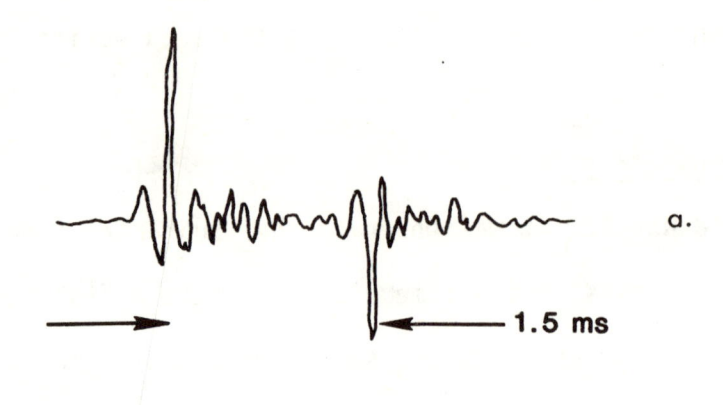

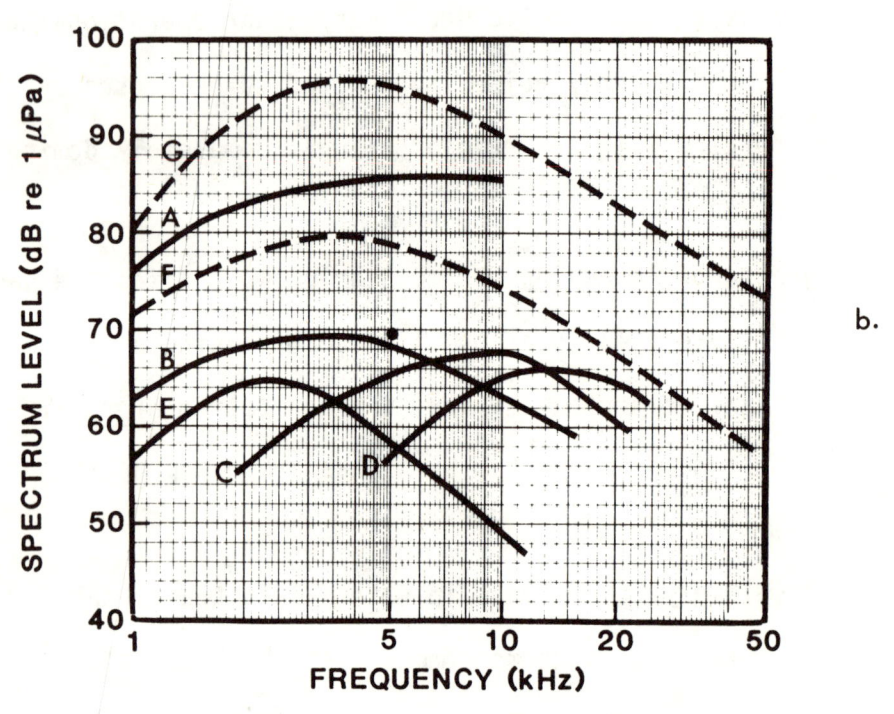

Figure 7-4. The noise of snapping shrimp. a: Oscillogram of a single shrimp snap; the second negative-going peak is the surface reflection, Reference 177. b: Spectra of shrimp noise. A: Near pilings in Kaneoke Bay, Oahu, Hawaii. B: Off Pt. Loma, San Diego. C: Florida-Bahamas area. D: Southwest Pacific. E: San Diego Yacht Harbor. F: Biscayne Key, off Miami. G: At a boat dock, water depth 10 ft, Miami. Curves A-E, Reference 177. Curves F and G, Reference 178.

A compilation of measured shrimp noise spectra is given in Figure 7-4b. The spectra evidently have a broad similarity of shape, but differ widely in level depending on the size and density of the shrimp beds and the distance to them from the measurement location.

7.3 Twenty-Cycle Pulses

When the first deep bottomed low-frequency hydrophones were installed in the early 1950's as part of Project Jezebel, which evolved into the present SOSUS (Sound Surveillance System), some peculiar noises were observed. These were trains of pulses of 20-Hz sound, each a few seconds long, that persisted for periods of minutes or hours. These pulses were so mysterious when first observed that their occurrence was placed in the SECRET category of security classification, wherein they remained for nearly ten years, and were given the unimaginative name of "twenty-cycle pulses." When it became clear that they were of non-human origin and were caused by some marine animal, they came to be described in the open literature, first by Walker (179), and have now been observed in many of the world's oceans.

Although characterized by a frequency closely centered near 20 Hz, the pulses occur in pulse trains having a variety of repetition patterns. The most common type found at Bermuda (179) had a time interval between equal-amplitude pulses of 12 sec. The next most common type was a doublet consisting of a large amplitude pulse followed 15 seconds later by a smaller amplitude pulse, the pair repeating 22 seconds later. A variety of other doublets was also observed at the Bermuda location. A distinct feature of these Bermuda signals was that when they were present for several hours they showed a sequence of continuous pulsing for about 15 minutes, and then stopped abruptly for about 2-1/2 minutes.

Twenty-cycle pulses, though with different repetitive characteristics, along with other low-frequency sounds, have been observed off the coast of southern California (180) (181) (182), the central Pacific (183), near Midway Island (184), off Hawaii (185), off New Zealand (186) and in the south Norwegian Sea (187).

Figure 7-5 is a sound spectrogram showing twenty-cycle doublets, each about 12 seconds long, in which Type I appears to be amplitude modulated and Type II is not.

The acoustic power of the pulses has been found to lie between 1.2 and 8 watts (188), corresponding to a source level between 173 and 181 dB re 1µPa at 1 yd. Triangulation by means of separated hydrophones showed that the source followed erratic tracks with speeds of a few knots. This evidence, together with visual sightings from aircraft and the regular occurrence of the signals during whale migration periods have shown that the twenty-cycle pulses are caused by one or more species of whale. But what species of whale, and by what process whales make this peculiar noise are still not entirely clear. Shevill, Watkins and Backus (189) have presented strong evidence that the pulses in the North Atlantic are produced by the common finback whale Balaenoptera physalus found in all oceans of the world. But the mechanism by which such a large amount of sound at such a low frequency can be produced is unknown. One suggestion is that they are the heartbeats of the whale, a hypothesis supported by the commonly occurring doublet characteristic of the pulses. Another hypothesis is that the pulses are associated in some way with the breathing mechanism, such as a lung resonance. In support of this hypothesis is an analysis (190) showing that a free 20-Hz resonant gas-filled sphere at a depth of 20 fathoms would have a radius of 30 cm, and, in order to radiate a power between 1 and 10 watts, would require a surface displacement

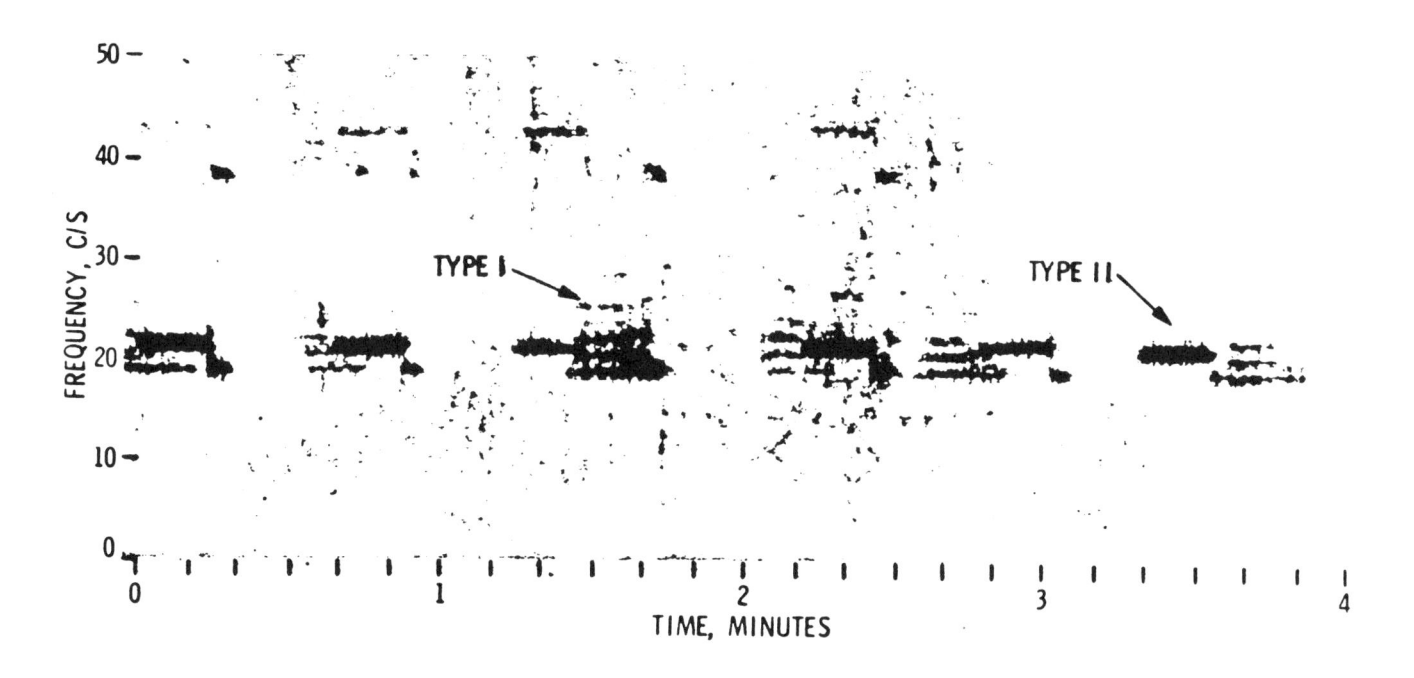

Figure 7-5. Sound spectrogram of long twenty-cycle pulses as recorded off San Clemente Island, California. The two types of pulses, one following the other, are distinguished by amplitude modulation. Reference 182.

amplitude between 0.05 and 0.11 cm. These amplitudes are not unreasonably large. Also in support of the breathing hypothesis is the fact that the pulse trains are interrupted, or disappear, at intervals corresponding to the breathing or surfacing cycles of the whale.

7.4 Non-Biologic Noises

In this category are a number of sources of diverse origin, ranging from the noise of falling rain to the noise caused by an underwater volcano. In this section, a few of these diverse noises will be described.

7.4.1 Rain

The study of noise made by the impact of water droplets and spheres falling on water has had, according to Franz (63), a long history dating back to 1919, and the mechanism of sound production appears to be well understood. Franz investigated the splash made by the gas-to-liquid entry of an object both in theory and in the laboratory by means of a hydrophone in a water-filled tank. It was found that an object hitting the air-water interface makes noise by three processes: (1) the impact itself, (2) the oscillations of the object after impact, and (3) the oscillations of the entrained air carried by the object below the surface and left behind as the object continues to fall. The first is said to dominate the others at high impact velocities, while the last is dominant at low velocities; the second is, of course, insignificant for rain. In this connection, Leslie (191) reported the noise made by bullets from a Wesson .38 caliber special revolver striking the water and found that most of the noise was caused by air bubble oscillations rather than by the impact.

The sound made by a spray--or rain of water droplets--depends upon the size and velocity of fall of the droplets and the rate or intensity of spraying. Franz obtained expressions for the spectrum level of spray noise in terms of these quantities. But, for rainfall, the droplet size and impact velocity are functions primarily of the single parameter--the rate of rainfall. Figure 7-6 shows the theoretical spectra of rain noise with rainfall rate as a parameter, along with two sets of measurements. Of particular note is the flatness of the spectra; within 3 dB, they are flat from 200 to 20,000 Hz, and are independent of the depth of the measurement hydrophone below the surface.

Field observations of rain noise in Long Island Sound have been reported by Heindsmann, Smith and Arneson (192). Their results are superposed on the theoretical spectra in the preceding figure. Although no measurements of rate of rainfall--except for the semi-quantitative categories of "heavy," "moderately heavy," and "intermittent"--were made, there is no more than general agreement between the observed and theoretical spectra.

On the other hand, the measurements of Bom (193), at the center of an Italian lake 250 m in diameter and of greatest depth 10 m, were accompanied by careful measurements of rainfall rate. The results were fitted by an expression of the form NL = A + B log R, where NL is the spectrum level at a specified frequency within the range 300 to 1600 Hz and R is the rain rate in inches per hour over the range 0.1 to 1 inch per hour. The quantities A and B were found to depend on frequency as follows:

Frequency Band	A	B
300 - 600 Hz	75.6	13.9
600 - 1200 Hz	74.5	13.8
1200 - 2400 Hz	74.0	14.7
2400 - 4800 Hz	74.1	16.3
4800 - 9600 Hz	71.9	16.1

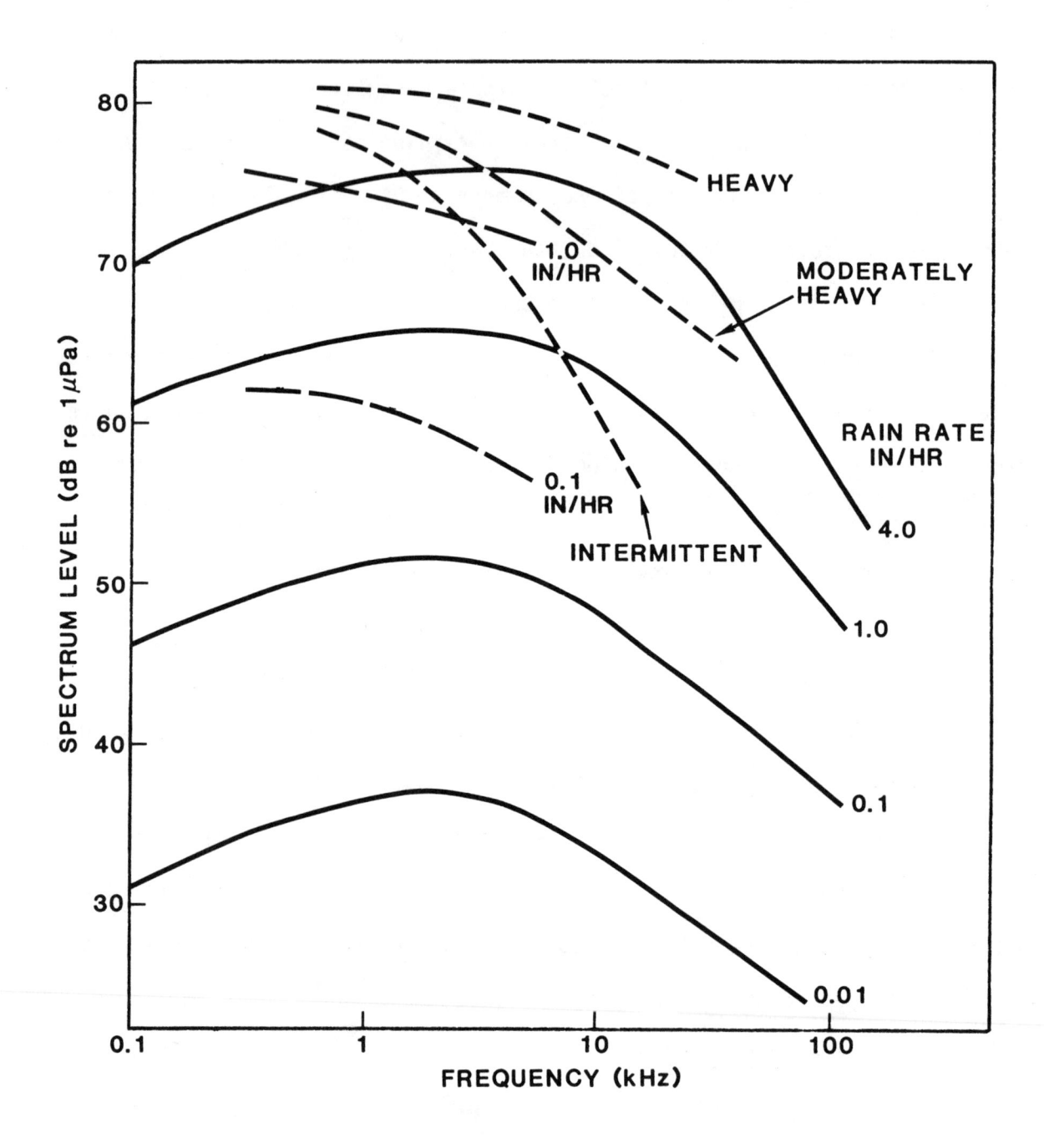

Figure 7-6. Theoretical rain noise spectra of Franz (solid curves) compared with two field measurements. Short dashes: data of Heindsman, Smith and Arneson (192). Long dashes: data of Bom (193).

The levels represented by the above expression are appreciably higher than those predicted by the theoretical spectra of Franz, even though the flatness of the spectra is roughly confirmed. Thus, at 1 kHz and a rainfall rate of 0.1 inch per hour, the comparison shows Franz at 51 dB and Bom at 60.7 dB. At 1 kHz and 1 inch per hour, the comparison shows Franz at 66 dB and Bom at 74.5 dB. The discrepancy is apparently too large to be explained by bottom reflections in the shallow lake in which the measured levels were obtained. Thus, while the noise-making processes of rain are apparently understood, the causes of the quantitative discrepancies between theory and field data evident in Figure 7-6 are still a mystery.

7.4.2 Earthquakes, Explosions and Volcanos

It has been noted above (see Section 2.2) that bottom unrest is a possible source of the ambient background of the sea at subsonic frequencies. An undersea earthquake, sometimes called a "seaquake," is a form of bottom unrest that stands out above the general prevailing background and can be identified as being caused by earth motion at some localizable spot in the earth. McGrath (21) placed a hydrophone on the bottom at the Mid-Atlantic Ridge along with a 2-component seismometer and observed over 500 seaquakes over a 7-day period. These were said to raise the broadband level over the range 5 to 32 Hz by at least 20 dB at this location of extreme seismic activity. In the Pacific, on the East Pacific Rise, sonobuoys were used by Northrop, Stixrud and Lovett (194) to successfully detect and measure microearthquakes. Also, a strong seaquake with the epicenter only 50 km away has been described (195) and recorded by a research vessel towing a hydrophone array for continuous seismic reflection profiling. Of course, in seismically quiet ocean areas only occasional strong earthquakes may be said to contribute to the noise background.

With the increasing use in recent years of explosives for offshore seismic exploration for oil, distant explosions have become common, and often annoying, sources of noise in the sea. For example, during a 12-month period beginning in August 1965, some 19,801 shots were identified (196) on recordings from hydrophones of the Missile Impact Location System (MILS) of the Pacific Missile Range. Most of these shots originated off the west coast of North America and north of Hawaii and were predominantly the result of distant offshore seismic exploration shooting. In the Gulf of Mexico at a deep-water location just south of the continental slope, impulsive seismic exploration noises could be detected audibly virtually all the time during a 12-hour recording period (36), and raised the levels by 20 to 30 dB during periods of intense activity. There seems to be no unclassified information on the noise made by one or more drilling oil wells in or near an oil field.

As a curiosity, we may mention the discovery and location of a submarine volcano by Northrop (197) in the western Pacific by means of recordings of a number of widely spaced MILS hydrophones. The signals were mostly below 100 Hz, with the major part of the energy in an octave band centered near 10 Hz. Thirteen other volcanoes in the Pacific and two in the Atlantic were said to have been monitored on MILS hydrophones during the years 1956-1974.

7.4.3 Surf

Very little scientific attention has been given to the contribution made by breaking surf to the ambient background in shallow water. Bardyshev (97) found significant departures from a normal distribution in the noise at ranges out to 600 m from a rocky coast, and attributed the spikiness to the breaking of waves and the impacts of rocks and pebbles on each other. Beyond a distance of 600 m, the noise was similar to that in the open sea. Wilson, Wolf and Ingenito (198) used directional DIFAR AN/SSQ-53A sonobuoys placed at different distances out to 8.5 km from the beach in Monterey Bay, California. During heavy surf conditions some 10 dB more noise was found at 8.5 km in 90 m of water in beams pointing toward shore than in beams pointing toward the open sea. This excess noise, presumably originating in the breaking surf, occurred principally in the frequency band 100 to 700 Hz; the lower frequency limit was doubtless the result of poor propagation in the very shallow water close to shore.

7.5 Summary

In this section we have described a variety of noises, of man-made and non-man-made origin that occur in the sea. These may be of only temporary duration and occasional occurrence at a particular location and so can hardly be said to be part of the "ambient" background at that location. On the other hand, they may be of such persistent and regular occurrence that they dominate the background of noise; examples are the noise of croakers in Chesapeake Bay and of snapping shrimp in shallow tropical waters. Such intermittent noises are often strong and cover a wide range of frequencies and, when they occur, they form a dominant and annoying background for the detection of signals in passive and active sonars.

8. NOISE IN THE ARCTIC

8.1 Introduction

Because of its ice cover, the Arctic is a unique noise environment, in which the noise depends upon the characteristics of the ice as well as upon wind speed and air temperature. Because of the variability of these factors, the levels of noise found in the Arctic cover a wide range, from about the levels of Knudsen Sea State 4 down to levels well below Sea State Zero--a level so low that ultra-low noise amplifiers in the receiving electronics are required.

Three summaries of Arctic noise have appeared. Two summaries, dated 1964 (199) and 1967 (200), are by A. R. Milne, the Canadian scientist who has done and written the most on the subject. The third summary (201), dated 1968, appears in the Russian literature and is based on 16 references, of which only two are to Russian work.

Strange to say, the acoustics of the Arctic did not receive attention until the late 1950's. At this time the nuclear-powered submarine came into being and made possible navigation under the ice, thereby providing a practical motivation for investigations of this peculiar underwater environment.

8.2 Sources of Noise in the Arctic

8.2.1 Ice Cracking

One source of noise peculiar to the Arctic is the cracking of the ice cover. This occurs in shore-fast spring and winter ice and consists of short bursts of noise originating in tensile cracks in the ice. These cracks occur near the surface of the ice and are the result of radiative cooling during periods of falling air temperatures. Surface cracking can occur only when the ice near the surface is brittle and exposed to the air with only a thin or nonexistent snow cover. The cracking bursts reach a maximum near midnight when the radiative cooling is a maximum. An example of the diurnal variability of this form of noise is shown in Figure 8-1, as reported by Milne for the month of April, 1968 (202). Each burst of cracking is roughly a decaying sinusoid lasting a few milliseconds, as illustrated by Figure 8-2.

The spectrum of this form of noise reaches a broad maximum in the decade 0.1 to 1.0 kHz, and increases during the nighttime hours as the ice continues to cool. Figure 8-3 shows the noise spectra at several times during the night as observed in McClure Strait by Milne and others (203) under shore-fast unbroken ice. There is evidently an increase of some 15 dB in the level in the band 0.1 to 1.0 kHz during the early nighttime hours.

These thermally induced tension cracks should be expected to occur only in relatively low-salinity, very cold sea ice where the ice will respond elastically to thermal stresses. Because the noise is made by individual cracks, the noise amplitudes are spiky and distinctly non-Gaussian, and have a Poisson distribution in time of occurrence. Figure 8-4 gives cumulative distribution curves for ice cracking sounds and the sound of rafting summer ice. In both cases, the noise is distinctly non-Gaussian. However, with increasing depth below the ice, the noise spikes become more numerous and the amplitudes acquire a quasi-Gaussian distribution, in accordance with the Central Limit Theorem.

8.2.2 The Wind

Under a solid ice cover when cracking noises are absent, the wind creates noise in the Arctic by its flow over the rough ice surface and through the impact of blowing snow. This form of noise has been found to have a flat spectrum at frequenies greater than 1 kHz, and is thereby in marked contrast with the spectrum of wind noise on the ice-free sea surface (which has a

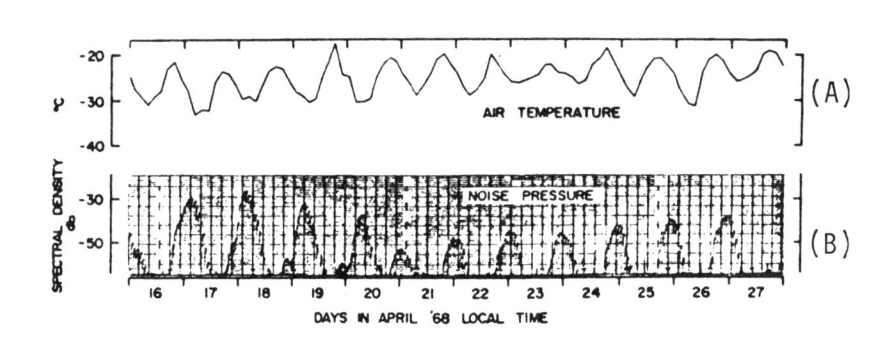

Figure 8-1. (A) Diurnal variations in air temperature in degrees C and (B) the noise level in the 150 to 300 Hz band. Reference 202.

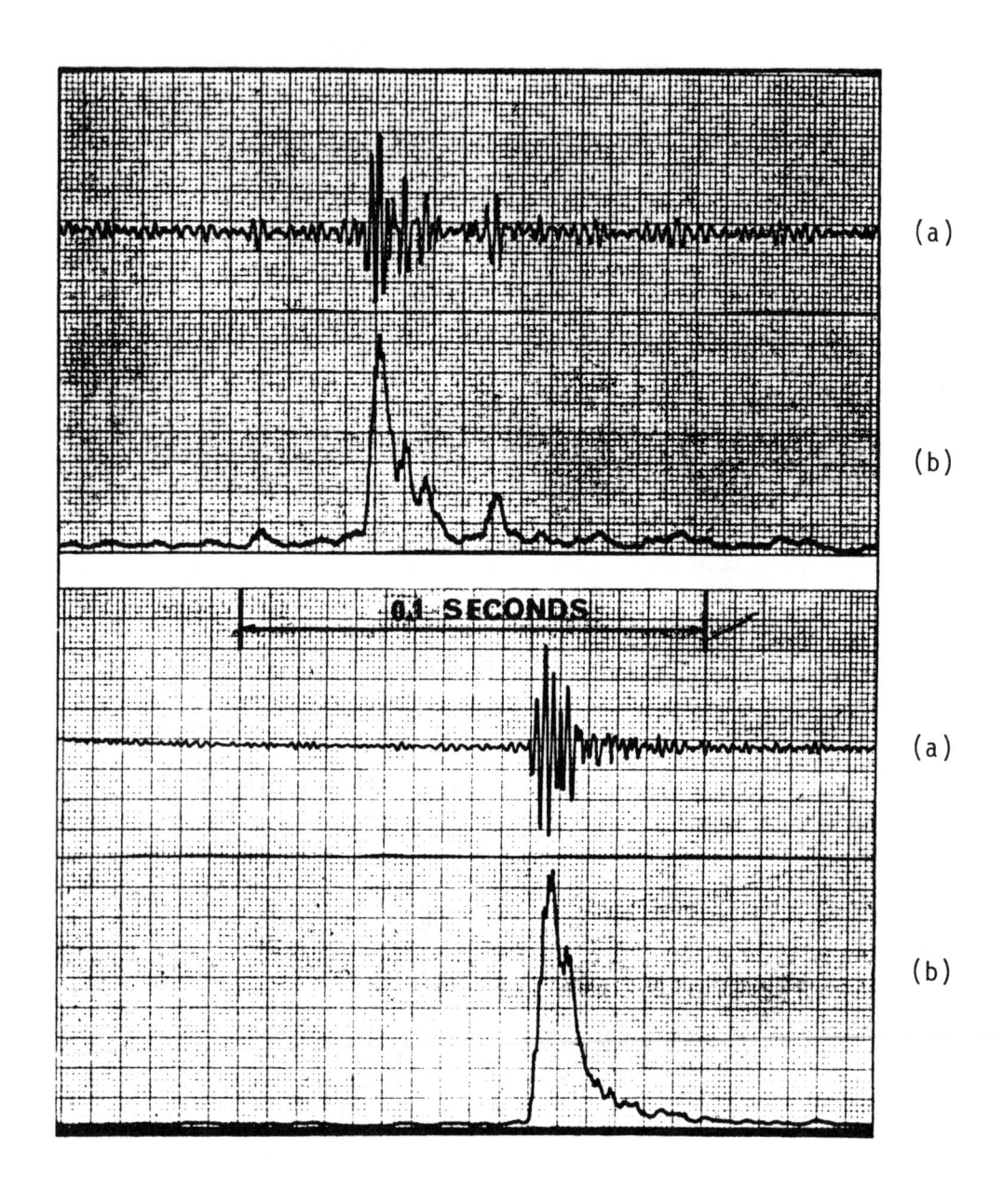

Figure 8-2. Ice-cracking pulses (a) and their envelopes (b), frequency band 200 to 600 Hz. Reference 200.

8-4

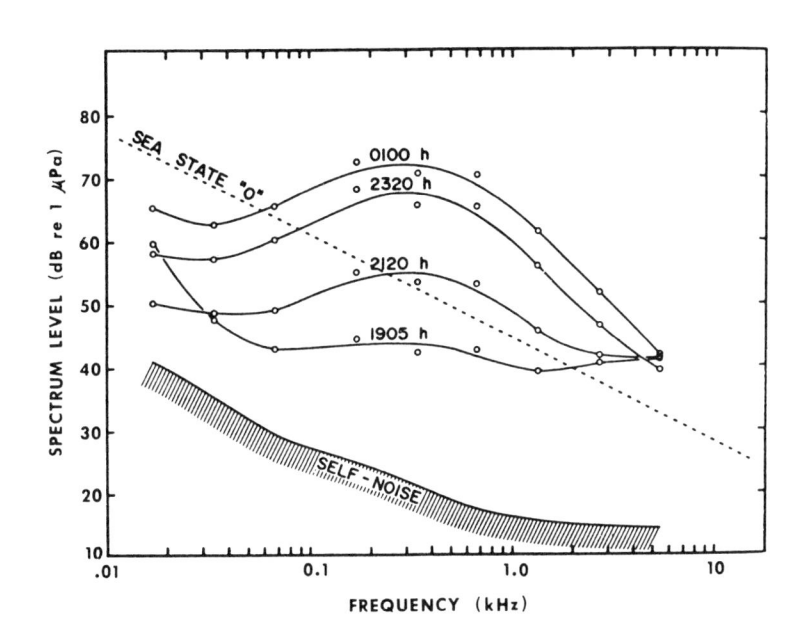

Figure 8-3. Spectra of ice-cracking noise during the nighttime hours. During the night, the air cools faster than the ice, producing a greater temperature difference and more noise than during the evening hours. Reference 203.

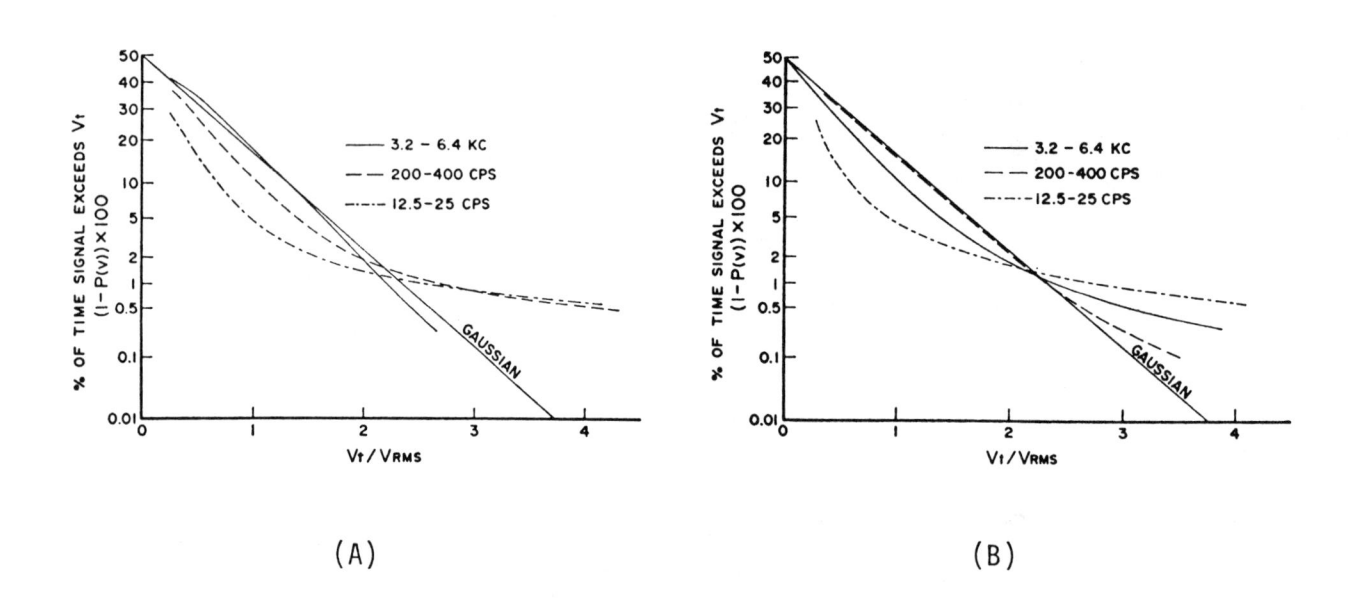

Figure 8-4. Cumulative distribution curves of the amplitude of (A) ice-cracking sounds in shore-fast springtime ice, and (B) sounds made by rafting summer ice, as observed in the Canadian Archipelago. Reference 199.

spectral slope of about 5 dB per octave). Figure 8-5 illustrates the spectrum of wind noise under a continuous ice cover for a number of wind speeds. Another contrasting feature is that the increase of noise with wind speed in the octave 3.2 to 6.4 kHz is faster over ice than over open water, as shown by Figure 8-6, where the under-ice noise increases at the rate of 16 dB per wind speed doubled in comparison with the value of 6 dB per wind speed doubled over the ice-free surface. Such differences as these suggest that the scale of turbulences responsible for the noise is different in the two cases (204). For example, it has been suggested by Milne (200) that the wind generates noise as it moves over ice by the collapse of the tiny eddies that form in the lee of tiny snow and ice crystals and thereby acquires a Gaussian amplitude distribution. The wind has also been postulated (86) to be the cause of continuous long-period vertical oscillations of the ice cover, principally in the period range 15 to 60 sec.

8.2.3 Moving Ice Masses

When the ice cover is not continuous but consists of summer ice floes, noise is made by the slow jostling of the ice blocks. This noise is said to be surprisingly low in level (200). On the other hand, the rafting and cracking of cold weather ice produces isolated noise bursts of extreme amplitudes that often overload the recording equipment (199). The amplitude distribution of summer ice noise is shown in Figure 8-4B, as observed in September under an ice cover of 70°/o one-year ice and 30°/o polar floes.

Off Cape North, Nova Scotia, in the Gulf of St. Lawrence, there are strong tidal currents causing significant movement of consolidated pack ice that never becomes shore-fast. The ice conditions can change markedly during a 24-hour period as a result of a changing wind direction. Noise observations by Oake (205) in 65 fathoms of water at the edge of the St. Lawrence Channel

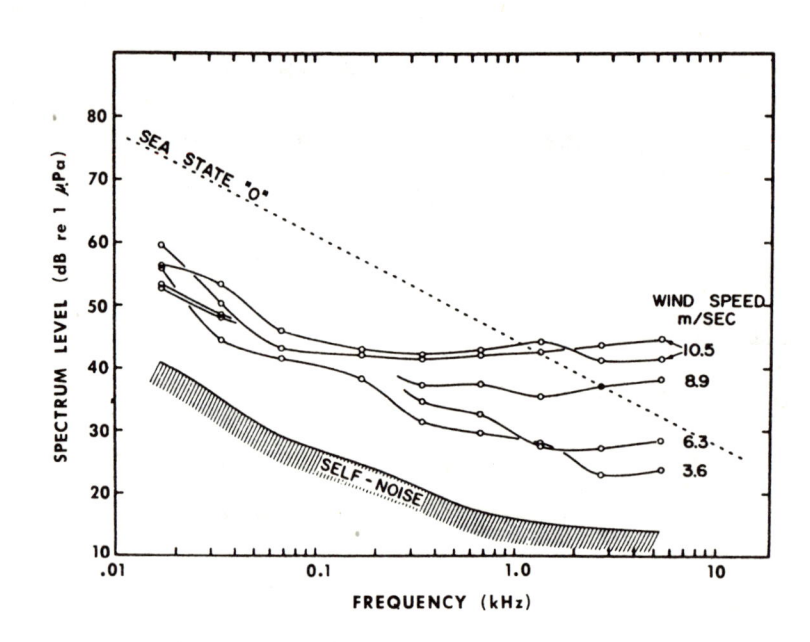

Figure 8-5. Spectra of wind noise under shore-fast ice for several wind speeds. Reference 203.

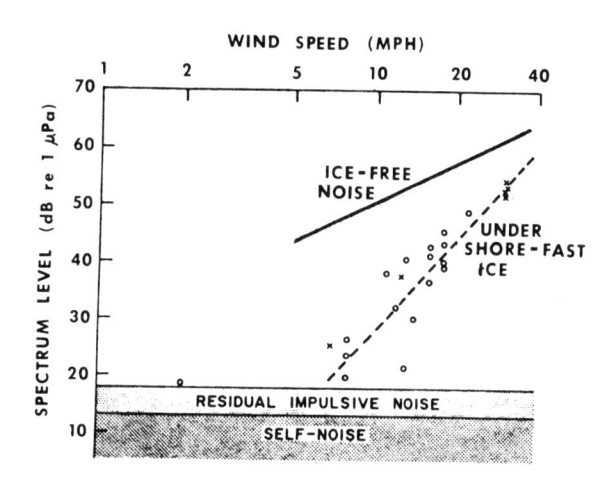

Figure 8-6. Wind noise levels under a continuous ice cover during the month of February in the frequency band 3.2 to 6.4 kHz. Reference 204.

showed wide variations lasting for periods of minutes due to the movement of ice floes along the shore. In addition, short bursts occurred as a result of shear stresses set up by the movement.

Another occurrence of intermittent strong noises was observed by Dwyer (206) with a hydrophone suspended in pack ice to a depth of 91 m. It picked up highly non-Gaussian noise containing bursts lasting from 0.1 to 0.3 seconds, caused by the rubbing and bumping of ice masses. These bursts were entirely different from the ice cracking transients of non-moving shore-fast ice. This sort of spikiness has also been noted by Greene and Buck (207) from observations on an ice floe in the Beaufort Basin.

Still another kind of transient noise is caused by ice ridging, where, in heavy ice conditions, ice floes are fractured and crushed, and ice blocks are piled up into ridges. In the wintertime central Beaufort Sea ridging is believed to be the single largest source of noise (224).

Noise measurements have been made during several winters from 1961 to 1964 at a site north of Prince Edward Island in the gulf of St. Lawrence and have been reported by Macpherson (208) and Payne (209) (210). At this location the noise background was found to decrease steadily from January to April as the ice cover increased from zero to several feet in thickness. At the same time the slope of the spectrum decreased and the noise level became less dependent upon wind speed as the ice cover became continuous by April, as shown by Figure 8-7 which presents noise level vs. wind speed, both with and without an ice cover, at a frequency of 141 Hz. However, this latter finding disagrees with the data obtained at a higher frequency (3.2 to 6.4 kHz) at a location further north (Lat. 78°), where a strong variation of noise with wind speed was observed (see Figure 8-6). In addition to this difference, no ice-cracking noises were reported at the Prince Edward Island site, probably because of the warmer and less brittle condition of the ice. In short, the interaction of the ice with the air above depends not only upon its roughness and snow cover, but also upon its temperature and the temperature of the air.

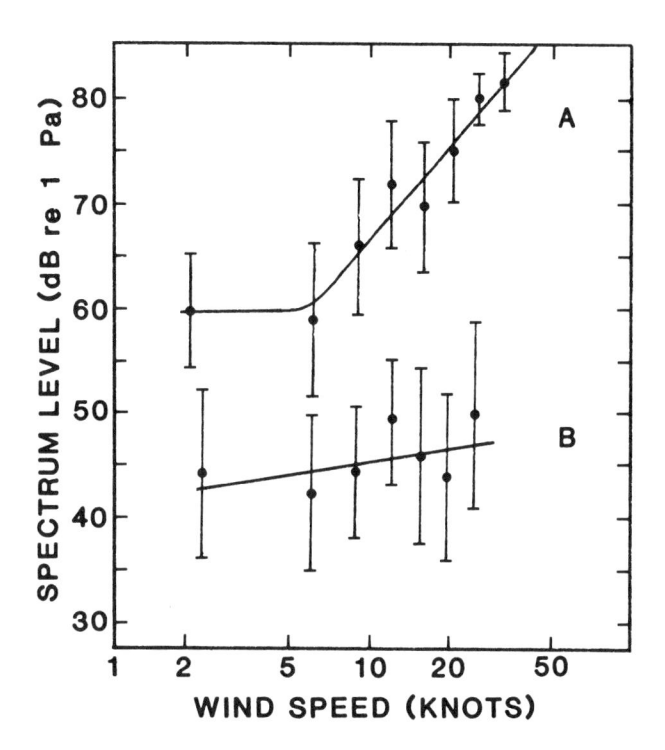

Figure 8-7. Noise level vs. wind speed at 141 Hz. A - open water, B - with a continuous ice cover (April), Prince Edward Island, Reference 209. The open water levels are the same as those found by Piggott (67) further east on the Scotian Shelf. The vertical bars show the standard deviation about the mean level.

8.3 The Marginal Ice Zone

The edge of an ice sheet is a source of noise that is caused by wave impact upon the ice floes at or near the edge of the ice. Near the boundary, the levels would be expected to be higher than those either in open water or well under the ice. This noise maximum at the ice edge was observed by Diachok and Winokur (211) by means of sonobuoys dropped at intervals across the ice-water boundary in the East Greenland Sea. Figure 8-8 shows the variation of level with distance each way, toward open water to the right and under the ice to the left, from a sharp compact ice edge. The maximum of noise at the edge was about 12 dB higher than the levels in open water and was about 20 dB higher than the levels inside the ice field. With a diffuse ice edge, the measured levels were 4 and 10 dB higher at the edge than in open water and far under the ice, respectively. Thus, the ice edge appears to act as a long-line noise source, producing a higher level in open water than under the ice at the same distance as a result of the poorer sound propagation under the ice. Similar results obtained in a similar way were reported by Peters (212).

In the Greenland Sea east of Greenland, two anchored hydrophones were used by McGrath (213) to obtain noise data over a period of time, one for several days, the other for nine months. During the longer recording period, the ice edge moved over the recording site, and in so doing provided confirming evidence for the conclusion that the ice edge acts as a strong source of noise, with levels dependent upon wind speed and direction relative to the edge. The highest levels occurred in early spring (March) and in the fall (November), and were 12 to 16 dB higher than the maximum values observed under a continuous ice cover.

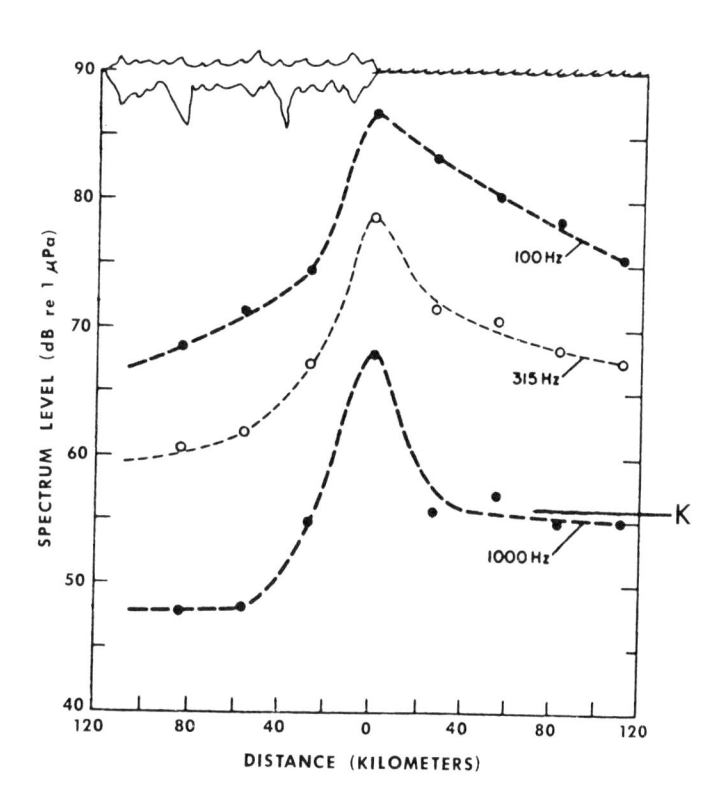

Figure 8-8. Noise level on either side of a sharp ice edge. Sea state 2. With a more diffuse ice edge, the peakedness of the curves is less pronounced. The line marked K is the level from the open water curves of Knudsen for the prevailing sea state. Reference 211.

8-13

Icebergs have been found by Urick (214) to make noise when they are melting. This noise appears to be caused by the escape of air contained under pressure in minute vacuoles in the ice, as a result of the melting ice wall. This kind of noise was found to have a flat or "white" spectrum extending out to a frequency at least as high as 10 kHz, and is so intense that it dominates the wind as a high-frequency source of noise in locations where actively melting icebergs are prevalent.

8.4 Biological Noise

Biological noises, often a dominant form of noise in ice-free lower latitudes, are insignificant in the Arctic. However, various drumming and pure-tone noises have been heard (200) that have been attributed to seals, but these noises do not have the level or time duration to be considered important contributors to the ambient background. On the other hand, in the Antarctic, in McMurdo Sound, biological noises of seals and humpback whales were strong enough to dominate the spectrum between 200 and 800 Hz, at least in the summer season (215).

8.5 Low-Frequency Noises

A number of peculiar sounds and vibrations occur at low frequencies in the ice-covered Arctic. For example, standing flexural gravity waves have been postulated to occur on individual ice floes. These waves generate pressures unattenuated with depth and occurring at twice the frequency of the standing wave in the water below, in the manner described theoretically by Longuet-Higgins (see Section 2.7). Low-frequency noise also can result from the buoyant bobbing of sea ice, which produces frequencies that depend on the thickness and density of the ice. This form of noise has been verified by

Milne (200) from sound spectrograms obtained under late summer ice consisting of 30°/o old disconnected polar floes and 70°/o one-year ice. Figure 8-9 shows broad tonal components at frequencies of 0.35 and 0.7 Hz, in agreement with the estimated thickness of the floes of one-year ice. Yet another form of low-frequency noise has been observed by Macpherson and Clark (216) at a frequency equal to the reciprocal of the round-trip travel time between surface and bottom, and its harmonics, under an ice cover in shallow water. These tonal components were attributed to the low-order propagation modes in the shallow water duct, but such modes appear not to have been reported for shallow ice-free water.

Underwater earthquakes, or "seaquakes", have been observed in the ice-covered Arctic, notably those originating along the mid-Arctic Ridge. A number of such quakes have been recorded by Keenen and Dyer (225) at a distance of 300 km from the Ridge. They have been found to last as long as 72 sec and to have a broad spectrum peaking near 5 Hz. These quakes not only travel through the earth's mantle to the vicinity of the receiving hydrophone, as most earthquakes do, but in the Arctic they are received more strongly as "T-waves". These waves are scattered by the ice in the vicinity of the quake, or epicenter, and they travel as waterborne sound under the Arctic ice cover.

8.6 Summary and Compilation

The quietest periods in the ice-covered Arctic occur with (1) a continuous shore-fast ice cover, (2) a rising air temperature and (3) a low wind speed. Under these conditions, the noise spectrum level is about 20 dB below Knudsen Sea State Zero, which is the lowest level occurring in the ice-free sea. On the other hand, when ice movement occurs, or when a continuous ice sheet cools rapidly, the noise level in the Arctic becomes some

20 dB greater than Knudsen Sea State Zero, so as to reach the levels of Sea State 3 or 4 in open water. Even higher levels occur near the edge of the ice in the Marginal Ice Zone.

In short, the ambient background depends upon the nature of the ice, whether continuous, broken, moving or shore-fast, the temperature of the air, and the speed of the wind.

Figure 8-10 is a compilation of reported spectra for the ice-covered Arctic including one for the Antarctic when biological noise existed. The spectra are keyed to the accompanying table wherein the area, a short description of conditions and the literature reference are given.

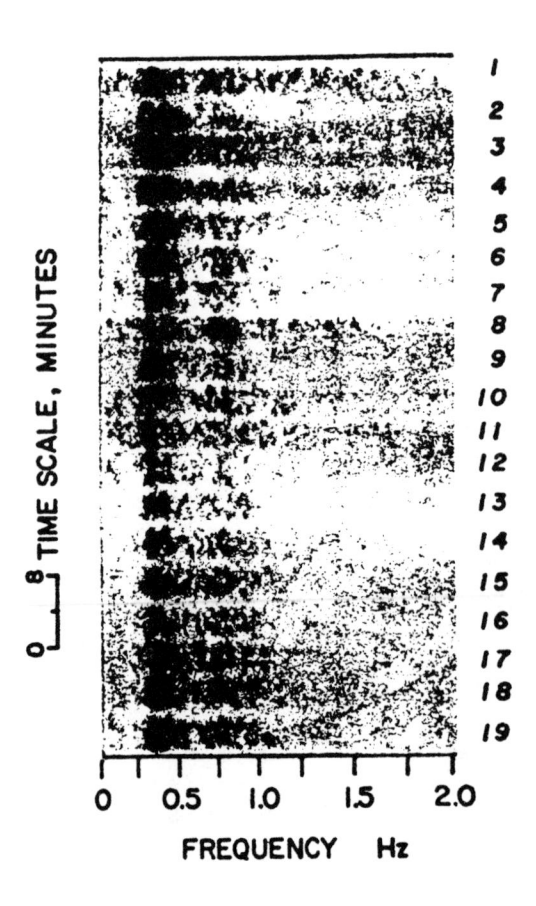

Figure 8-9. Sound spectrograms showing the low frequency buoyant bobbing frequencies of late summer ice floes. The data was obtained in September 1961 by Milne in the McClure Strait. Reference 200.

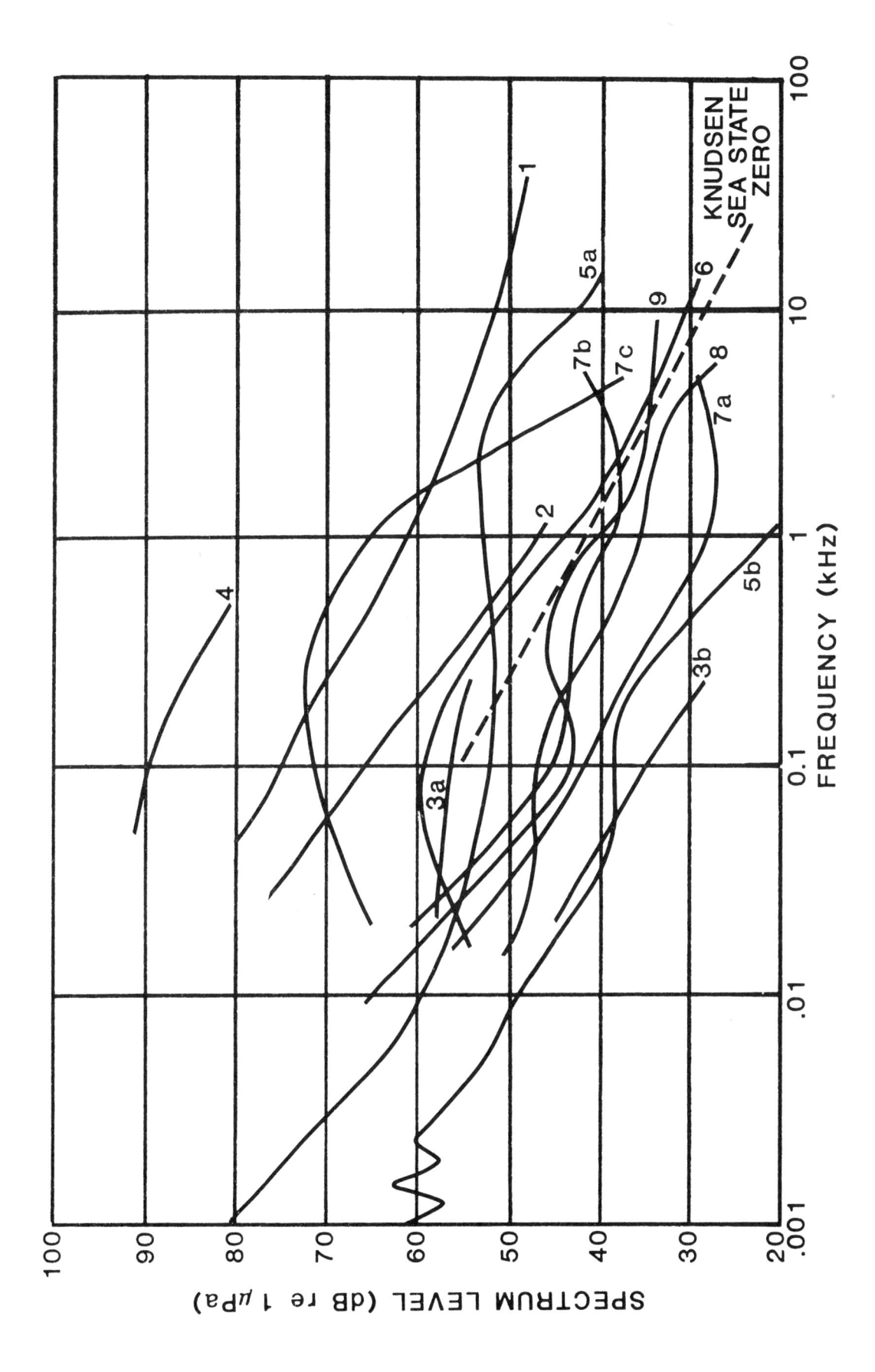

Figure 8-10. A compilation of noise spectra in the Arctic. The conditions for each spectrum are listed in the accompanying table.

TABLE OF LOCATIONS AND ICE CONDITIONS

(Adjunct to Figure 8-10)

Spectrum No.	Location	Ice Conditions	Reference
1	Cape North, Nova Scotia	Moving pack ice, solid cover with on-shore winds	205
2	Beaufort Basin	On an ice floe	207
3	Gulf of St. Lamere, N. of P.E. Island	a) January, noise due to ice collisions b) April, consolidated ice	208
4	East Greenland Sea	Marginal Ice Zone	213
5	Canadian Archipelago	8 to 12 ft solid ice, April a) Cracking noises strong b) Cracking noises absent	199
6	Canadian Archipelago	70°/o 1-year ice, 30°/o floes, September, wind less than 15 kn	199
7	McClure Strait	Shore-fast ice, May a) Warming periods b) Cooling, 1905 hours c) Cooling, 0100 hours	203
8	Canada Deep	10/10 springtime pack ice, wind speed 4.4 kn	203
9	McMurdo Sound, Antarctica	Wind less than 1 kn, biological noise present	215

9. STATE OF PRESENT KNOWLEDGE

Ambient noise is one of the most satisfying aspects of underwater sound in that it is relatively well understood. Based upon roughly 2000 published reports and papers (though many, however, are trivial and classified) we now have noise models from which we can make a rough prediction of noise level at a specified frequency for an area in which noise measurements are not available, based on estimates of wind speed, ship traffic, and biologic content for the area of interest. Wilson (226) has described a noise model for calculating the noise level to be expected, assuming a given distribution of storm winds and shipping, using a suitable propagation model. In a subsequent paper (227) Wilson described a modification appropriate for a hand calculator. Agreement was obtained between observations and model predictions for a Pacific Ocean location. It may be said that the existence and validity of such models as these is evidence of the excellence of our understanding of the causes of ambient noise in the sea. More than that, the Naval Underwater Systems Center has published (228) a set of standard values and curves, that, once adopted by sonar analysts and design engineers, will provide, at least for the time being, a consistent set of data for modeling and design applications.

Still, there are a number of topics concerning ambient noise that need further investigation. More measurements and studies are needed at frequencies below 20 Hz, which is the region of the spectrum containing the fundamental blade-rate frequency and lower harmonics of propeller-driven vessels. At the other end of the frequency scale, there is little or no real-world data above 20 kHz, where we must now rely on theory plus extrapolation of lower-frequency data. Another uncertain subject is the process by which the wind makes the noise in the sea; several processes have been postulated. Further studies are needed to find the relationship between wind stress and noise in different regions of the spectrum.

Some more minor subjects for future research are (1) the noise of falling rain, to clear up the discrepanices in existing data (see Figure 7-6), (2) the horizontal coherence and directionality of surface-generated noise, (3) more and valid field data to verify the theoretical dependence of surface noise with depth (see Figure 4-1) at a high enough frequency to assure a lack of contamination by shipping noise, and (4) the process of generation of the twenty-cycle pulses by whales.

It is a truism to say that prediction is the sole end product of research in an environment that cannot be modified or controlled. Two examples are weather prediction in the air and earthquake prediction in the earth. For ambient noise, any future measurements should be preceded by predictions of what to expect, based upon previous measurements in an analogous area and estimates of the environmental parameters affecting the noise. For ambient noise in the sea, the prediction problem involves using estimated environmental data for the area of interest, together with existing measurements of noise in a similar area, based on the understanding as expressed by our current noise models.

APPENDIX

IDEALIZED AMBIENT NOISE SPECTRA BASED ON THE LITERATURE

Idealized noise spectra in the range 1 to 1000 Hz are presented in Figure A. These eyeball spectra are based upon (1) a sampling of measured data and (2) conjecture and guesswork, relying on our knowledge of the spectral behavior of the sources of noise.

The data points as plotted were read off from various figures included in the present report. Each data point is identified as to a key number in Table A, plus the wind speed in knots or a short dash if wind speed was not applicable or stated. Thus, 5/20 means a value read off from item number 5 (see Figure 2-17A) at 20 kn; 3/- refers to item 3 (see Figure 2-8) where the wind speed was irrelevant or not stated.

The selected frequencies were 1, 10, 20, 50, 200 and 1000 Hz. Above 1000 Hz the spectra can be extrapolated at the Knudsen rate of -5 dB per octave or -18 dB per decade. Below 10 Hz only a single solid line is drawn because of lack of data at the present time, even though it is likely that wind-dependent noise exists in this frequency range.

On inspection, the data points are seen to be widely inconsistent from report to report, doubtless as a result of differences in area, water and receiver depth, averaging employed, and other conditions. For this reason the curves should be considered to be merely eyeball estimates based on the disparate plotted data points. Various "intermittent" noises, such as the "twenty-cycle pulses" and rain noise, are excluded.

Yet these generalized spectra may be said to have some practical value whenever a grand average level, valid worldwide, is required, as in a calculation using the sonar equations for general conditions. For a specific

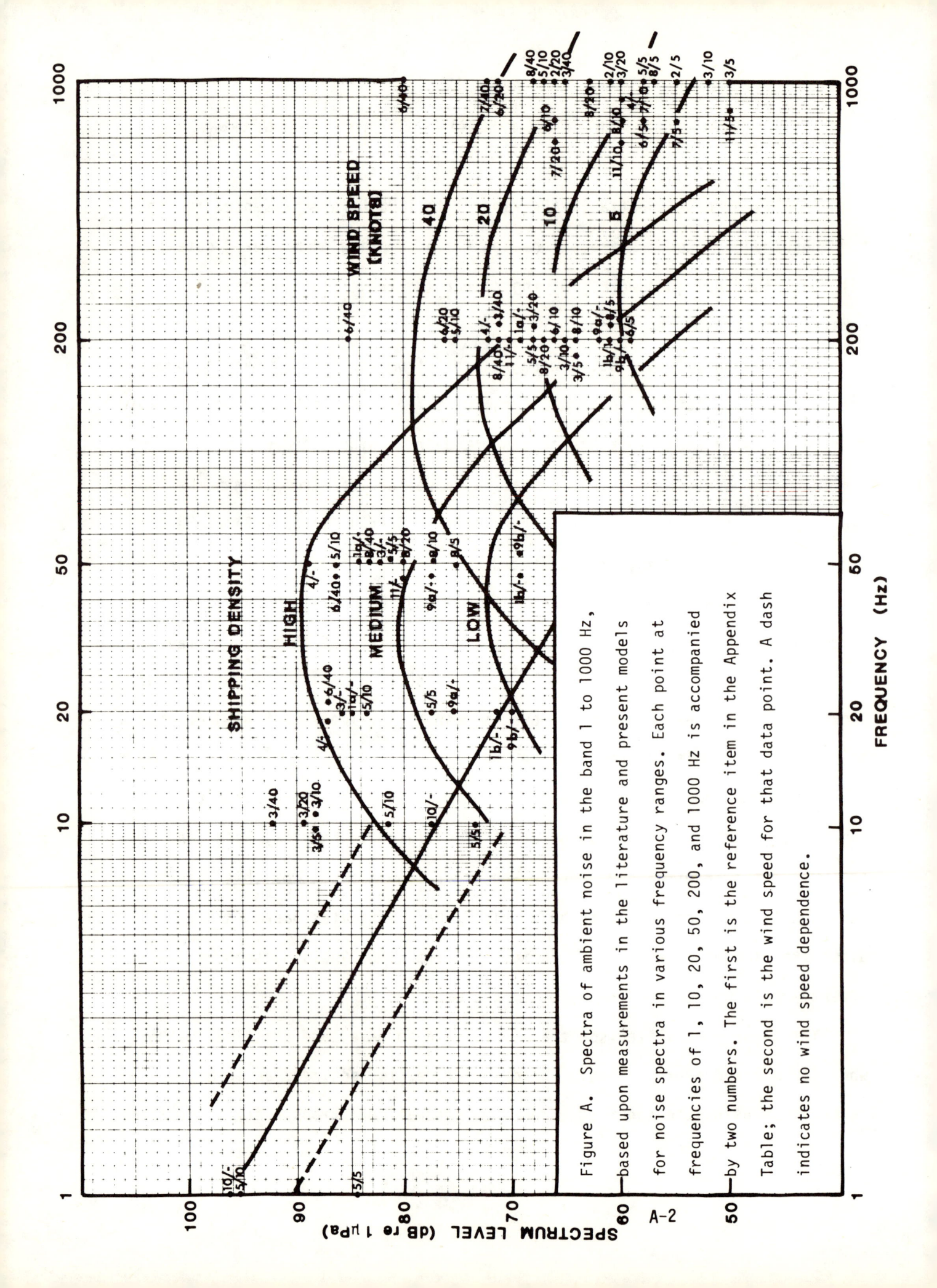

Figure A. Spectra of ambient noise in the band 1 to 1000 Hz, based upon measurements in the literature and present models for noise spectra in various frequency ranges. Each point at frequencies of 1, 10, 20, 50, 200, and 1000 Hz is accompanied by two numbers. The first is the reference item in the Appendix Table; the second is the wind speed for that data point. A dash indicates no wind speed dependence.

TABLE A

Key Number	Figure Number	Location and Condition
1a	2-6	North Pacific
1b	2-6	South Pacific
2	2-10	Knudsen curves, 1000 Hz only
3	2-8	Northwest Atlantic
4	2-9	Mediterranean, average of 4 stations, 1000-Hz points extrapolated from 800 Hz
5	2-17A	Russian data, location unknown, 200 fm
6	2-17B	On Scotian Shelf, depth 28 fm
7	3-7	Compilation of shallow water levels at 1 kHz
8	2-13	Off Australia and New Zealand, 1000 Hz extrapolated from 800 Hz
9a	2-14	West of Guam, on the bottom at 4572 m
9b	2-14	West of Guam, 1000 m above bottom
10	2-3	Average of several published spectra in the range 1 to 10 Hz
11	2-12	North of St. Croix, water depth 13,000 ft, 1000 Hz interpolated between 500 and 3150 Hz

area, the literature should be consulted. The reader is invited to draw his own set of canonical curves based on the data in Figure A plus any other data that may be available.

LIST OF REFERENCES*

1. Knudsen, V. O., R. S. Alford and J. W. Emling, Survey of Underwater Sound Report No. 3: Ambient Noise, Office of Scientific Research and Development Report 4333, Sect. No. 6.1 - NDRC-1848, 1944. Also, Underwater Ambient Noise, J. Mar. Res. 7, 410, 1948.

2. Urick, R. J. and A. W. Pryce, A Summary of Underwater Acoustic Data, Part V, Background Noise, Office of Naval Research, 1954.

3. Wagstaff, R. A., A Comprehensive Ambient Noise Bibliography, Naval Undersea Center, TP 333, 1973.

4. Gold, B. A. and J. X. Palumbo, A Computerized Ambient Noise Bibliography, Acoustic Environmental Support Detachment Tech. Note 73-01, 1973.

5. Ingalsbee, T. E., The Ambient Noise Bibliographic Data Bank, Acoustic Environmental Support Detachment Tech. Note 70-03, 1975, AD B006108L.

6. Palumbo, J. X. and B. A. Gold, A Computerized Ambient Noise Bibliography, Updated Version, Naval Oceanographic Office Tech. Note 6160-1-74, AD 918741L.

7. Gold, B. A. World-Wide Distribution of Published Ambient Noise Data, Naval Oceanographic Office TN-6160-9-73, 1973, AD 768740.

8. Wenz, G. M., Acoustic Ambient Noise in the Ocean: Spectra and Sources, JASA 34, 1936, 1962.

9. Wenz, G. M., Acoustic Ambient Noise in the Ocean: Supplementary Remarks, JUA 13, 517, 1963.

10. Arase, T. and E. M. Arase, A Review of Recent Research in Underwater Ambient Noise, JUA 15, 589, 1965.

11. Crouch, W. W., Ambient Sea Noise: A Review of the Literature, Naval Underwater Systems Center Tech. Rep. 4179, 1972, AD 893161.

12. Wenz, G. M., Review of Underwater Acoustics Research: Noise, JASA 51, 1010, 1972.

---

*The following abbreviations are used: JASA, Journal of the Acoustical Society of America; JUA, U.S. Navy Journal of Underwater Acoustics; SPA, Soviet Physics (Acoustics). Reports with AD numbers may be obtained from the National Technical Information Service (NTIS), U.S. Department of Commerce, Springfield, VA 22151.

13. Furduev, A. V., Underwater Noise of Dynamical Origin, SPA 9, 219, 1964.

14. Furduev, A. V., Sea Noise, Joint Publications Research Service Rep. No. 65187, 1975. Available through NTIS.

15. <u>Handbook of Oceanographic Tables</u>, U.S. Naval Oceanographic Office, 1966.

16. Nichols, R. H., Infrasonic Ambient Ocean Noise Measurements: Eleuthera, JASA 69, 974, 1981.

17. Daniels, F. B., Mechanisms of Generation of Infrasound by Ocean Waves, JASA 24, 83, 1952.

18. Daniels, F. B., Generation of Infrasound by Ocean Waves, JASA 34, 352, 1962.

19. Urick, R. J., Sea-Bed Motion as a Source of the Ambient Noise Background of the Sea, JASA 56, 1010, 1974.

20. Schneider, W. A. and M. M. Backus, Ocean-Bottom Seismic Measurements off the California Coast, J. Geophys. Res. 69, 1135, 1964.

21. McGrath, J. R., Infrasonic Sea Noise at the Mid-Atlantic Ridge Near 37°N, JASA 60, 1290, 1976.

22. Kibblewhite, A. C., J. A. Shooter and S. L. Watkins, Examination of Attenuation at Very Low Frequencies Using the Deep Water Ambient Noise Field, JASA 60, 1040, 1976.

23. Perrone, A. J., Infrasonic and Low-Frequency Ambient Noise Measurements on the Grand Banks, JASA 55, 754, 1974.

24. Strasberg, M., Nonacoustic Noise Interference in Measurements of Infrasonic Noise, JASA 66, 1487, 1979.

25. McGrath, J. R., O. M. Griffin and R. A. Finger, Infrasonic Flow-Noise Measurements Using an H-58 Omnidirectional Cylindrical Hydrophone, JASA 60, 390, 1977.

26. Bardyshev, V. I., A. M. Velikanov and S. G. Gershman, Experimental Studies of Underwater Noise in the Ocean, SPA 16, 512, 1971.

27. Buck, B. M. and C. R. Greene, A Two-Hydrophone Method of Eliminating the Effects of Non-Acoustic Noise Interference in Measurements of Infrasonic Ambient Noise Levels, JASA 68, 1306, 1980.

28. Lomask, M. R. and Saenger, R. A., Ambient Noise in a Deep Inland Lake, JASA 32, 878, 1960.

29. Perrone, A. J., Deep-Ocean Ambient Noise Spectra in the Northwest Atlantic, JASA 46, 762, 1969.

30. Cato, D. H., Ambient Sea Noise in Waters Near Australia, JASA 60, 320, 1976.

31. Bannister, R. W. and others, Measurements of the Low-Frequency Wind Generated Ambient Noise in the Deep Ocean, Naval Underwater Systems Center Tech. Doc. 6565, 1981, AD A105748.

32. Burgess, A. S. and D. J. Kewley, Wind-Generated Surface Noise Source Levels in Deep Water East of Australia, JASA 73, 201, 1983.

33. King, L. A., Ambient Noise in Shipping Channels, paper presented at Acoustic Communication Workshop, Washington, D.C., August 1982.

34. Arase, E. M. and T. Arase, Ambient Noise in the Mediterranean Sea, Hudson Labs., Columbia University Tech. Rep. 152, 1968, AD 392736.

35. Wilson, J. H., Site and Frequency Dependence of Ambient Noise in the Northeastern Pacific Ocean, JASA 73, 539, 1983.

36. Shooter, J. H., T. E. DeMary and R. A. Koch, Ambient Noise in the Western Gulf of Mexico, Applied Res. Labs., University of Texas, ARL-TR-82-15, 1982.

37. Knudsen, V. O., R. S. Alford and J. W. Emling, Underwater Ambient Noise, J. Mar. Res. 7, 410, 1948.

38. Johnson, R. L., Ambient Noise Measurements off the West Coast of Mexico and Central America, U.S. Navy Electronics Lab. Rep. 474, 1954, AD 761619.

39. Perrone, A. J., Correlation of Deep Ocean Ambient Noise Spectra with Wind Speed and Wave Height for Frequencies from 11 to 2816 Hertz, Underwater Sound Lab. Rep. 935, 1968, AD 843764.

40. Urick, R. J., G. R. Lund and T. J. Tulko, Depth Profile of Ambient Noise in the Deep Sea North of St. Croix, Virgin Islands, Naval Ordnance Lab., NOLTR 73-176, 1972.

41. Shooter, J. A. and M. L. Gentry, Wind-Generated Noise in the Parece Vela Basin, JASA 70, 1757, 1981.

42. Crouch, W. W. and P. J. Burt, Logarithmic Dependence of Surface-Generated Sea Noise Spectrum Load on Wind Speed, JASA 51, 1066, 1972. Also, Naval Underwater Systems Center Tech. Memo LA 11-5-71, 1971, AD A066772.

43. Anon., Ocean Ambient Noise Used to Measure Wind Speed, JASA 68, 1221, 1980.

44. Shaw, P. T., D. R. Watts and H. T. Rossby, On the Estimation of Oceanic Wind Speed and Stress from Ambient Noise Measurements, Deep-Sea Research 25, 1225, 1978.

45. Mellen, R. H., Thermal-Noise Limit in the Detection of Underwater Acoustic Signals, JASA 24, 478, 1952.

46. Sullivan, E. J. and K. A. Kemp, Some Notes on "Thermal Noise Limit in Detection of Underwater Acoustic Signals," (by R. Mellen, JASA 24, 478, 1952), JASA 70, 631, 1981.

47. Ezrow, D. H., Measurement of the Thermal-Noise Spectrum of Water, JASA 34, 550, 1962.

48. Anderson, A. L. and G. J. Gruber, Ambient-Noise Measurements at 30, 90 and 150 kHz in Five Ports, JASA 49, 928, 1971.

49. Li, H., On Wind Induced Underwater Ambient Noise, Naval Ocean Research and Development Activity Tech. Note 89, 1981, AD A101-682.

50. Isakovich, M. A. and B. F. Kuryanov, Theory of Low Frequency Noise in the Ocean, SPA 16, 49, 1970.

51. Wilson, J. H., Very Low Frequency Wind-Generated Noise Produced by Turbulent Pressure Fluctuations in the Atmosphere Near the Ocean Surface, JASA 66, 1499, 1979.

52. Kuo, E. Y., Deep-Sea Noise Due to Surface Motion, JASA 43, 1017, 1968.

53. Yen, N. and A. J. Perrone, Mechanisms and Modeling of Wind Induced Low-Frequency Ambient Sea Noise, Naval Underwater Systems Center Tech. Rep. 5833, 1979.

54. Longuet-Higgins, M. S., A Theory of the Origins of Microseisms, Trans. Roy. Soc. (London) A, 243, 1, 1950.

55. Marsh, H. W., Origin of the Knudsen Spectra, JASA 35, 409, 1963.

56. Brekhovskikh, L. M., Underwater Sound Waves Generated by Surface Waves in the Ocean, Atmospheric and Oceanic Physics (Izvestiya) 2, 970, 1966.

57. Harper, E. Y. and P. G. Simpkins, On the Generation of Sound in the Ocean by Surface Waves, J. Sound and Vibration 37, 185, 1974.

58. Hughes, B., Estimates of Underwater Sound (and Infrasound) Produced by Non-Linearly Interacting Ocean Waves, JASA 60, 1032, 1976.

59. Lloyd, S. P., Underwater Sound from Surface Waves According to the Lighthill-Ribner Theory, JASA 69, 425, 1981.

60. Brekhovskikh, L. M., Acoustics and the Ocean, SPA 24, 361, 1978.

61. Furduev, A. V., Undersurface Cavitation as a Source of Noise in the Ocean, Akademia Nauk SSR Izvestia, Atmospheric and Oceanic Physics 2, 523, 1966.

62. Wilson, J. H., Low Frequency Wind-Generated Noise Produced by the Impact of Spray with the Ocean Surface, JASA 68, 952, 1980.

63. Franz, G. J., Splashes as Sources of Sound in Liquids, JASA 31, 1080, 1959.

64. Morris, G. B., Depth Dependence of Ambient Noise in the Northeastern Pacific Ocean, JASA 64, 581, 1978.

65. Kerman, B. R., Underwater Sound Generation by Breaking Waves, JASA 75, 148, 1984.

66. Masterov, E. R. and S. P. Shorokhova, Results of an Experimental Study of the Energy Spectral Characteristics of Sea Noise, SPA 19, 139, 1973.

67. Piggott, C. L., Ambient Sea Noise at Low Frequencies in Shallow Water of the Scotian Shelf, JASA 36, 2152, 1964.

68. Wylie, D. V., Sea Noise Measurements in the Coral, Solomon and Bismark Seas, Australian Weapons Res. Estab. Tech. Memo 432, 1971, AD B040403.

69. Browning, D. G., N. P. Fisch and P. D. Koenigs, Shallow Water Ambient Noise: Offshore Measurements at 20-10,000 Hz, Naval Underwater Systems Center Tech. Rep. 6825, 1982, AD B070162L.

70. Arase, E. M. and T. Arase, Ambient Sea Noise in the Deep and Shallow Ocean, JASA 42, 73, 1967.

71. Bardyshev, V. I. and others, Wind Speed Dependence of the Undersea Noise Level, SPA 17, 252, 1971.

72. Penhallow, W. S. and F. T. Dietz, Correlation of 630-cps Shallow Water Ambient Noise with Wind Speed and Wavelength, JASA 36, 2149, 1964.

73. Wagstaff, R. A., Low Frequency Ambient Noise in the Deep Sound Channel-- The Missing Component, JASA 69, 1009, 1981.

74. Wales, S. C. and O. I. Diachok, Ambient Noise Vertical Directionality in the Northwest Atlantic, JASA 70, 577, 1981.

75. Northrop, J., M. S. Loughridge and E. W. Werner, Effect of Near-Source Bottom Conditions on Long-Range Sound Propagation in the Ocean, J. Geophys. Res. 73, 3905, 1968.

76. Worley, R. D. and R. A. Walker, Low Frequency Ambient Noise and Sound Transmission Over a Thinly Sedimented Rock Bottom, JASA 71, 863, 1982.

77. Willis, J. and F. T. Dietz, Some Characteristics of 25 cps Shallow Water Ambient Noise, JASA 37, 125, 1965.

78. Calderon, M. A. Probability Density Analysis of Ocean Ambient and Ship Noise, U.S. Navy Electronics Lab. Rep. 1248, 1964.

79. Arase, E. M. and T. Arase, On the Statistics of Ambient Noise, Hudson Laboratories, Columbia University Informal Document 126, 1963, AD A074105.

80. Urick, R. J., Models for the Amplitude Fluctuations of Narrow-Band Signals and Noise in the Sea, JASA 62, 878, 1977.

81. Bannister, R. W. and others, Variability of Low Frequency Ambient Sea Noise, JASA 65, 1156, 1979.

82. Urick, R. J., G. R. Lund and T. J. Tulko, St. Croix Revisited: Further Measurements of the Ambient Acoustic Background at Different Depths at a Deep Sea Location, Naval Ordnance Lab., NOLTR 73-31, 1973.

83. Gray, L. M. and D. S. Greeley, Source Level Model for Propeller Blade Rate Radiation for the World's Merchant Fleet, JASA 67, 516, 1980.

84. Hecht, R. J. and L. A. Mole, Acoustic Signal Processing Extensions, JASA 64, 1206, 1978.

85. Perrone, A. J. and L. A. King, Analysis Technique for Classifying Wind- and Ship-Generated Noise Characteristics, JASA 58, 1186, 1975.

86. Dyer, I., Statistics of Distant Shipping Noise, JASA 53, 564, 1973.

87. Mikhalevsky, P. N. and I. Dyer, Approximations to Distant Shipping Noise Statistics, JASA 63, 732, 1978.

88. Hunkins, K., Waves on the Arctic Ocean, Jour. Geophys. Res. 67, 2477, 1962.

89. Wenz, G. M., Low-Frequency Deep-Water Ambient Noise Near Bermuda, JUA 24, 243, 1974. Also, Ambient Noise Spectrum Levels as a Function of Water Depth, JASA 48, 362, 1970.

90. Perrone, A. J., Summary of a One-year Ambient Noise Measurement Program off Bermuda, Naval Underwater Systems Center Tech. Rep. 4979, 1976, AD A024462.

91. Walkinshaw, H. M., Low Frequency Spectrum of Deep Ocean Ambient Noise, JASA 32, 1497, 1960 (Abstract only).

92. Ross, D., *Mechanics of Underwater Noise*, Pergamon Press, New York, 1976, p. 285.

93. Urick, R. J., The Underwater Acoustic Environment at Two Contrasting Shallow Water Locations, Naval Ordnance Lab., TR 71-69, 1971.

94. Urick, R. J. and D. L. Bradley, Ambient Acoustic Environment of Shallow Water off Ft. Lauderdale, Fla., Naval Ordnance Lab. Tech. Rep. 70-72, 1970.

95. Weston, D. E., Mode Colouration in Shallow Water Ambient Noise, Admiralty Research Lab. Rep. ARL/L/R86, 1970.

96. Urick, R. J., G. R. Lund and D. L. Bradley, Observations of Fluctation of Transmitted Sound in Shallow Water, JASA 45, 683, 1969.

97. Bardyshev, V. I., N. G. Kozhelupova and V. I. Kryshnii, Study of Underwater Noise Distributions in Inland Sea and Coastal Regions, SPA 19, 95, 1973.

98. Wenz, G. M., Some Periodic Variations in Low-Frequency Acoustic Ambient Noise Levels in the Ocean, JASA 33, 64, 1961.

99. Nichols, R. H. and C. E. Sayer, Frequency-Frequency Correlations of Ocean Ambient Noise levels, JASA 61, 1188, 1977.

100. Anderson, V. C., Nonstationary and Non-Uniform Oceanic Background in a High-Gain Acoustic Array, JASA 67, 1170, 1980.

101. Hodgkiss, W. S. and V. C. Anderson, Detection of Sinusoids in Ocean Acoustic Background Noise, JASA 67, 214, 1980.

102. Arase, T. and E. M. Arase, Deep-Sea Ambient Noise Statistics, JASA 44, 1679, 1968.

103. Jobst, W. J. and S. L. Adams, Statistical Analysis of Ambient Noise, JASA 62, 63, 1977.

104. Urick, R. J., Some Directional Properties of Deep-Water Ambient Noise, Naval Research Laboratory Rep. 3796, 1951.

105. Bradner, H. and A. Parvulescu, Attenuation of Surface-Generated Noise Received Deep in the Ocean, JASA 62, 1037, 1977.

106. Abramowitz, M. and I. A. Stegun, eds., Handbook of Mathematical Functions, National Bureau of Standards Applied Math. Ser. 55, 1965, p. 227.

107. Wagstaff, R. A., Ambient Noise Depth Dependence, JUA 23, 543, 1973.

108. Weston, D. E., Ambient Noise Depth Dependence Models and Their Relation to Low-Frequency Attenuation, JASA 67, 530, 1980.

109. Cavanagh, R. C. and W. R. Renner, Vertical Directionality and Depth Dependence of Averaged Signals and Noise, JASA 68, 1467, 1980.

110. Kibblewhite, A. C. and others, Factors Controlling the Ambient Noise Field Below the Deep Sound Channel, JUA 27, 551, 1977.

111. Anderson, A. L. and K. C. Focke, Model Sensivitity Studies: Ambient Noise Depth Dependence Related to Propagation Loss, JUA 28, 219, 1978.

112. Anon., Ambient Sea Noise Measured from 100 to 2240 Meter Depths at Locations Approximately 220 Nautical Miles from San Diego, Naval Undersea Center Tech. Note 650, 1971, AD A067137.

113. Perrone, A. J., Ambient Noise Levels as a Function of Water Depth, JASA 48, 362, 1970.

114. Lomask, M. and R. Frassetto, Acoustic Measurements in Deep Water Using the Bathyscaph, JASA 32, 1028, 1960.

115. Kibblewhite, A. C., G. E. Ellis and L. D. Hampton, An Examination of the Deep Water Ambient Noise Field in the Northeast Pacific Ocean, JUA 27, 373, 1977.

116. Urick, R. J., Ambient Noise in the Surface Duct, JUA 26, 79, 1976.

117. Urick, R. J., *Principles of Underwater Sound*, 3rd ed., McGraw-Hill Book Co., New York, 1983, p. 151.

118. Anderson, A., R. Martin and S. Marshall, Effects of Topographic Blockage and Ocean Boundaries on Low Frequency Noise Fields, paper in Proceedings of a Conference on Underwater Ambient Noise, SACLANTCEN, 11-14 May 1982, AD A125324.

119. Morris, G. B., Preliminary Results on Seamount and Continental Slope Reflection Enhancement of Shipping Noise, Marine Physical Lab., Scripps Inst. of Oceanogr. Ref. 75-34, 1975, AD A023476.

120. Talham, R. J., Ambient Sea-Noise Model, JASA 36, 1541, 1964.

121. Fox, G. R., Ambient Noise Directivity Measurements, JASA 36, 1537, 1964.

122. Becken, B. A., Directional Distribution of Ambient Noise in the Ocean, Scripps Inst. of Oceanogr. Rep. 61-4, 1961.

123. Axelrod, E. H., B. A. Schoomer and W. A. VonWinkle, Vertical Directionality of Ambient Noise in the Deep Ocean at a Site Near Bermuda, JASA 37, 77, 1965.

124. Stone, J., Problems Associated with the Measurement of Ambient Noise Directivity by Means of Linear Additive Arrays, JASA 34, 328, 1962.

125. Anderson, M. W. and R. L. Tittle, Analysis of an Integral Arising from the Investigation of Vertical Directivity of Sea Noise, JASA 45, 1129, 1969.

126. Yen, N., Ambient Sea Noise Directionality: Measurement and Processing, Naval Underwater Systems Center Tech. Rep. 5545, 1977, AD A038461.

127. Anderson, V. C., Arrays for the Investigation of Ambient Noise in the Ocean, JASA 30, 470, 1958.

128. Rudnick, P. and E. D. Squier, Fluctuations and Directionality in Ambient Sea Noise, JASA 41, 1347, 1967.

129. Forster, C. A., Ambient Sea Noise Directivity, paper presented at the 64th Meeting of the Acoustical Society of America, November, 1962.

130. Atkins, G. B., informal communication, Naval Air Development Center, 1974.

131. Browning, D. G. and others, Vertical Directionality of Low-Frequency Ambient Noise in the South Fiji Basin, Naval Underwater Systems Center Tech. Doc. 6611, 1982, AD A110500.

132. Anderson, V. C., Variation of the Vertical Directionality of Noise with Depth in the North Pacific, JASA 66, 1446, 1979.

133. Wagstaff, R. A., Interactive Technique for Ambient-Noise Horizontal-Directionality Estimation from Towed Line-Array Data, JASA 63, 863, 1978.

134. Wilson, J. H., Spatial Correlation of Wind Noise at Very Low Frequencies, JASA 60, 315, 1976.

135. Wagstaff, R. A., Horizontal Directionality Estimation Considering Array Tilt and Noise Field Vertical Arrival Structure, JASA 67, 1287, 1980.

136. Wagstaff, R. A. and J. W. Aitkenhead, Horizontal Directionality of Ambient Noise in the SOFAR Channel of the Northeast Pacific Ocean, JASA 26, 279, 1976.

137. O'Connor, J. C. P. and others, Horizontal Directionality of Ocean Ambient Noise, Bolt Beranek and Newman, Inc., Report 2728, 1974, AD A002247.

138. Yen, N., Ambient-Sea-Noise Directionality: Measurement and Processing, JASA 62, 1176, 1977.

139. Cron, B. F. and R. L. Shaffer, Array Gain for the Case of Directional Noise, JASA 41, 864, 1967.

140. Jacobson, M. J., Space-Time Correlation in Spherical and Circular Noise Fields, JASA 34, 971, 1962.

141. Cron, B. F. and C. H. Sherman, Spatial Correlation Functions for Various Noise Models, JASA 34, 1732, 1962.

142. Cox, H., Spatial Correlations in Arbitary Noise Fields with Application to Ambient Sea Noise, JASA 54, 1289, 1973.

143. Liggett, W. S. and M. J. Jacobson, Noise Covariance and Vertical Directivity in a Deep Ocean, JASA 39, 280, 1966.

144. Edie, J., Space-Time Correlation Functions of Surface-Generated Noise, Litton Systems Inc., TR 63-9-Bf, 1963.

145. Urick, R. J., Correlative Properties of Ambient Noise at Bermuda, JASA 40, 1108, 1966.

146. Lytle, D. W. and P. H. Moose, Space-Time Correlation Functions for a Surface Noise Model, JASA 39, 587, 1966.

147. Rein, C. R., Vertical Cross Correlation of Noise from a Horizontal Surface, JASA 50, 321, 1971.

148. Liggett, W. S. and M. J. Jacobson, Covariance of Noise in Attenuating Media, JASA 36, 1183, 1964.

149. Liggett, W. S. and M. J. Jacobson, Covariance of Surface-Generated Noise in a Deep Ocean, JASA 38, 303, 1965.

150. Buckingham, M. J., Array Gain of a Broadside Vertical Array in Shallow Water, JASA 65, 148, 1979.

151. Buckingham, M. J., A Theoretical Model of Ambient Noise in a Low-Loss, Shallow Water Channel, JASA 67, 1186, 1980.

152. Buckingham, M. J., Spatial Coherence of Wind Generated Noise in a Shallow Ocean Channel, JASA 70, 1412, 1981.

153. Kuperman, W. A. and F. Ingenito, Spatial Correlation of Surface Generated Noise in a Stratified Ocean, JASA 67, 1988, 1980.

154. Lloyd, L. S. and M. J. Daintith, An Analysis of High Frequency Ambient Noise, NATO SACLANT ASW Research Centre Memo SM-159, 1982, AD A125308.

155. Arase, E. M. and T. Arase, Correlation of Ambient Noise in the Ocean, JASA 33, 146, 1965.

156. Arase, E. M. and T. Arase, Mapping of the Space-Time Correlation of Ambient Sea Noise, JASA 40, 499, 1960.

157. Arase, E. M. and T. Arase, Correlation of Ambient Sea Noise, JASA 40, 205, 1966.

158. Cron, B. F., B. C. Hassell and F. S. Keltonic, Comparison of Theoretical and Experimental Values of Spatial Correlation, JASA 37, 523, 1965.

159. Urick, R. J., Coherence of Ambient Noise and the Signal from a Steady Source at Different Depths at a Deep Sea Location, Naval Ordnance Lab., NOLTR 73-68, 1973.

160. Linnette, H. M. and R. J. Thompson, Directivity Study of the Noise Field in the Ocean, Employing a Correlative Dipole, JASA 36, 1788, 1964.

161. Backus, R. H., Sound Production by Marine Animals, JUA 8, 191, 1958.

162. Tavolga, W. N., ed., *Marine Bioacoustics*, Pergamon Press, 1964.

163. Fish, M. P. and W. H. Mowbray, *Sounds of Western North American Fishes*, Johns Hopkins University Press, Baltimore, MD, 1970.

164. Schevill, W. E., Underwater Sounds of Cetaceans, in *Marine Bioacoustics*, W. N. Tavolga, ed., Pergamon Press, New York, 1964, p. 307.

165. Kaznadzei, V. V., S. A. Kreichi and E. N. Khakhalkina, Types of Dolphin Communication Signals and Their Organization, SPA 22, 484, 1976.

166. Cummings, W. C. and P. O. Thompson, Underwater Sounds from the Blue Whale, Balaenoptera Musculus, JASA 50, 1193, 1971.

167. Au, W., R. W. Floyd and J. E. Haun, Propagation of Atlantic Bottlenose Dolphin Echolocation Signals, JASA 64, 411, 1978.

168. Corcella, A. T. and M. Green, Investigation of Impulsive Deep-Sea Noise Resembling Sounds Produced by a Whale, JASA 44, 483, 1968.

169. Fish, J. F. and C. W. Turl, Acoustic Source Levels of Four Species of Small Whales, Naval Undersea Center TP 547, 1976, AD A037620.

170. Loye, D. P. and Proudfoot, D. A., Underwater Noise Due to Marine Life, JASA 18, 446, 1946.

171. Fish, M. P., Biological Sources of Sustained Ambient Sea Noise, in Marine Bioacoustics, W. N. Tavolga, ed., Pergamon Press, New York, 1964, p. 175.

172. Cato, D. H., Some Unusual Sounds of Apparent Biological Origin Responsible for Sustained Background Noise in the Timor Sea, JASA 68, 1056, 1980.

173. Cato, D. H., Marine Biological Choruses Observed in Tropical Waters Near Australia, JASA 64, 736, 1978.

174. Clapp, G. A., Periodic Variations of the Ambient Noise Level of Biological Origin off Southern California, U.S. Navy Electronics Lab. Tech. Memo TM-1027, 1966, AD A048863.

175. Fish, J. F. and W. C. Cummings, A 50 dB Increase in Sustained Ambient Noise from Fish (Cynoscion Xanthulus), JASA 52, 1266, 1972.

176. Painter, D. W., Ambient Noise in a Coastal Lagoon, JASA 35, 1458, 1963.

177. Everest, F. A., R. W. Young and M. W. Johnson, Acoustical Characteristics of Noise Produced by Snapping Shrimp, JASA 20, 137, 1948.

178. Widener, M. W., Ambient Noise Levels in Selected Shallow Water off Miami, Florida, JASA 42, 904, 1967.

179. Walker, R. A., Some Widespread High-Level Underwater Noise Pulses of Apparent Biological Origin off Cape Code, in Marine Bioacoustics, W. N. Tavolga, ed., Pergamon Press, New York, 1964, p. 121. Also, JASA 35, 1816, 1963.

180. Calderon, M., Low Frequency Underwater Pulsed Noise Observed in the Pacific Ocean, JUA 12, 729, 1962.

181. Calderon, M., Low Frequency Pulses Observed in the San Nicolas Basin, U.S. Navy Electronics Lab., TM-784, 1965.

182. Thompson, P. O., Marine Biologic Sounds West of San Clemente Island, U.S. Navy Electronics Lab. Rep. 1290, 1965.

183. Northrop, J., W. O. Cummings and P. O. Thompson, 20-Hz Signals Observed in the Central Pacific, JASA 43, 383, 1968.

184. Northrop, J., W. C. Cummings and M. F. Morrison, Underwater 20 Hz Signals Recorded Near Midway Island, JASA 49, 1909, 1971.

185. Patterson, B., P. H. Grayston and G. R. Hamilton, Movement of the Unknown 20-cps Source through Bermuda Waters, JUA 12, 489, 1962.

186. Kibblewhite, A. C., R. N. Denham and D. J. Barnes, Unusual Low-Frequency Signals Observed in New Zealand Waters, JASA 41, 644, 1967.

187. Weston, D. E. and R. I. Black, Some Unusual Low Frequency Biological Noises Underwater, Deep-Sea Res. 12, 295, 1965.

188. Patterson, B. and G. R. Hamilton, Repetitive 20-Cycle Biological Hydroacoustic Signals at Bermuda, in Marine Bioacoustics, W. H. Tavolga, ed., Pergamon Press, 1964, p. 121.

189. Schevill, W. E., W. A. Watkins and R. H. Backus, The 20-Cycle Signals and Balaenoptera (Fin Whales), in Marine Bioacoustics, W. N. Tavolga, ed., Pergamon Press, 1964, p. 147.

190. Harris, G. G., Physics of Sound Production by Fishes, in Marine Bioacoustics, W. H. Tavolga, ed., Pergamon Press, New York, 1964, p. 233.

191. Leslie, C. B., Underwater Noise Produced by Bullet Entry, JASA 36, 1138, 1964.

192. Heindsman, T. E., R. H. Smith and A. D. Armeson, Effect of Rain upon Underwater Noise Levels, JASA 27, 378, 1955.

193. Bom, N., Effect of Rain on Underwater Noise Level, JASA 45, 150, 1969.

194. Northrop, J., T. E. Stixrud and J. R. Lovett, Sonobuoy measurements of Seaquakes on the East Pacific Rise Near San Benedicto Island, Deep-Sea Res. 23, 519, 1976.

195. Birch, F. S., An Earthquake Recorded at Sea, Bull. Seism. Soc. Am. 56, 361, 1966.

196. Spiess, F. N., J. Northrop and E. W. Werner, Location and Enumeration of Underwater Explosions in the North Pacific, JASA 43, 640, 1968.

197. Northrop, J., Detection of Low-Frequency Underwater Sounds from a Submarine Volcano in the Western Pacific, JASA 56, 837, 1974.

198. Wilson, O. B., S. N. Wolf and F. Ingenito, Measurements of Acoustic Ambient Noise in Shallow Water Due to Breaking Surf, Naval Postgraduate School, NPS 61-83-006, 1982.

199. Milne, A. R. and J. H. Ganton, Ambient Noise Under Arctic Sea Ice, JASA 36, 855, 1964.

200. Milne, A. R., Sound Propagation and Ambient Noise Under Sea Ice, paper in Underwater Acoustics, Vol. 2, V. M. Albers, ed., Plenum Press, NY, 1967, pp. 120-138.

201. Bogorodskii, V. A. and Gusev, A. V., Under-Ice Noise in the Ocean, SPA 14, 127, 1968.

202. Milne, A. R., Thermal Tension Cracking in Sea Ice: A Source of Under-Ice Noise, J. Geophys., Res. 77, 2177, 1972.

203. Milne, A. R., J. H. Ganton and D. J. McMillin, Ambient Noise Under Sea Ice and Further Measurements of Wind and Temperature Dependence, JASA 41, 525, 1967.

204. Ganton, J. H. and A. R. Milne, Temperature and Wind-Dependent Ambient Noise Under Midwinter Pack Ice, JASA 38, 406, 1965.

205. Oake, A. M., Collection and Preliminary Analysis of Under-Ice Ambient Noise Data, Cape North, January to April 1971, Master's Thesis, Naval Postgraduate School, 1971, AD 733177.

206. Dwyer, R. F., FRAM II Single Channel Ambient Noise Statistics, Naval Underwater Systems Center Tech. Doc. 6583, 1981, AD A108755.

207. Greene, C. R. and B. Buck, Arctic Ocean Ambient Noise, JASA 36, 1218, 1964.

208. Macpherson, J. D., Some Under-Ice Acoustic Ambient Noise Measurements, JASA 34, 1149, 1962.

209. Payne, F. A., Effect of Ice-Cover on Shallow Water Ambient Sea Noise, JASA 36, 1943, 1964.

210. Payne, F. A., Further Measurements on the Effect of Ice Cover on Shallow-Water Ambient Sea Noise, JASA 41, 1374, 1967.

211. Diachok, O. I. and R. S. Winokur, Spatial Variability of Underwater Ambient Noise at the Arctic Ice-Water Boundary, JASA 55, 750, 1974.

212. Peters, K. W., Ambient Noise Levels in the Marginal Sea-Ice Zone East of Greenland, Naval Postgraduate School, Master's Thesis, 1979.

213. McGrath, J. R., Depth and Seasonal Dependence of Ambient Sea Noise Near the Marginal Ice Zone of the Greenland Sea, Naval Research Laboratory Rep. 7819, 1976, AD A025212.

214. Urick, R. J., The Noise of Melting Icebergs, JASA 50, 337, 1971.

215. Kibblewhite, A. C. and D. A. Jones, Ambient Noise Under Antarctic Sea Ice, JASA 59, 790, 1976.

216. Macpherson, J. D. and S. R. Clark, Resonances in Seismic Noise Under Arctic Sea Ice, Bull. Seism. Soc. Am. 54, 1797, 1964.

217. Milne, A. R., Statistical Description of Noise Under Shore-Fast Ice in Winter, JASA 39, 1174, 1966.

218. Urick, R. J. and G. R. Lund, Low-Frequency Coherence of Long Range Explosive Sounds and Ambient Noise in the Deep Sea, Naval Ordinance Lab. Rep. 70-111, 1970.

219. Revie, J. and D. E. Weston, Hydrophone Signals Due to Tidal and Wave Effects, Deep-Sea Res. 18, 545, 1971.

220. Talpey, T. E. and R. D. Worley, Infrasonic Ambient Noise Measurements in Deep Atlantic Water, JASA 75, 621, 1984.

221. Wille, P. C. and D. Geyer, Measurements on (sic) the Wind-dependent Ambient Noise Variability in Shallow Water, JASA 75, 173, 1984.

222. Dashen, R. and W. Munk, Three Models of Global Ocean Noise, JASA 76, 540, 1984.

223. Sloboda, R. S. and M. A. Manness, Spatial Coherence in Semicircular Noise Fields, JASA 74, 1260, 1983.

224. Pritchard, R. S., Arctic Ocean Background Noise Caused by Ridging of Sea Ice, JASA 75, 419, 1984.

225. Keenan, R. E. and I. Dyer, Noise from Arctic Ocean Earthquakes, JASA 75, 819, 1984.

226. Wilson, J. H., Wind-generated Noise Modeling, JASA 73, 211, 1983.

227. Wilson, J. H., Distant Storm Noise versus Local Wind Noise at 165 Hz in the Northeastern Pacific Ocean, JASA 74, 1500, 1983.

228. Sadowski, W., R. Katz and K. McFadden, Ambient Noise Standards for Acoustic Modeling and Analysis, Naval Underwater Systems Center Tech. Doc. 7265, 1984.

INDEX

Absorption, effect on depth variation, 4-1

Ambient noise, defined, 1-1

Anemometer based on noise level, 2-27

Arctic noise:
   biological noise, 8-14
   compilation of data, 8-15
   ice cracking, 8-1
   low frequency noises, 8-14
   wind noise, 8-2

Australia and New Zealand, noise near, 2-14, 2-19, 3-4, 5-9, 7-7

AUTOBUOY, 5-9

*****

Bathymetry effects, 2-41, 4-12

Beaufort wind force vs sea state, 1-5

Bermuda, noise at, 2-14, 2-34, 3-9, 4-4, 5-6, 6-2, 6-8, 6-10, 7-11

Bibliographies, 1-3

Biologic noise, 7-1

Bristol Channel, 3-16

*****

Cable strumming, 2-11

Coherence:
   and arrays, 6-1
   to distinguish pseudo-noise, 2-13
   theory of, 6-2

Correlation coefficent, defined, 6-1

Croakers, 7-6

Deep Sound Channel (DSC), 2-37

Directionality of noise:
   depth dependence, 5-13
   in the horizontal, 5-15
   theory of, 5-1
   in the vertical, 5-4

Down-slope conversion, 2-37, 4-14

*****

Earthquakes, 7-17

English Channel, noise in, 2-14, 4-8

*****

FLIP, 4-8, 5-15

Flow-noise, 2-8

Fluctuation spectrum, 3-1

Fort Lauderdale, Florida, noise at, 3-12, 3-16

*****

Ice cracking noises, 8-1

*****

Greenland Sea, noise in, 8-12

Grand Banks, noise near, 3-10

Gulf of Maine, noise in, 2-39, 3-12

Gulf of Mexico, noise in, 2-16

*****

History of research, 1-1

*****

Line components in noise, 2-1, 3-4

*****

I-1

Marginal ice zone, 8-12

Mediterranean Sea, noise in, 2-16, 4-5

Megaphone effect, 2-39

Microbaroms, 2-2

*****

Narragansett Bay, noise in, 2-37

*****

Pacific Ocean, noise in, 2-8, 2-16, 2-19, 2-23, 2-34, 3-4, 4-8, 5-15 7-17

Pseudo-noise, 2-8

Propagation effects on noise, 2-37

Pyroelectric noise, 2-2

*****

Reference unit, 1-4

Rain, 7-14

*****

St. Croix, noise at, 2-19, 3-4, 4-5, 6-14

Sea bottom motion, 2-2

Sea surface, processes of noise generation, 2-30

Seasonal effects, 3-10

Sea state vs wind speed, 1-5

Seismic exploration, 2-16

Shallow water, noise in, 2-18, 2-28, 3-12, 7-9

Standard deviation of noise samples:
  measurements of, 3-4
  in narrow bands, 3-4
  of shipping noise, 3-7

Snapping shrimp, 7-9

SONODIVER, 4-4

Sources of noise:
  high sonic band
    (200-50,000 Hz), 2-18
  infrasonic band (1-20 Hz), 2-4
  low sonic band (20-200 Hz), 2-13
  ultra-low frequency band
    (< 1 Hz), 2-1
  ultrasonic band (> 50 kHz), 2-27

SPARBUOY, 4-4

Spray and cavitation, 2-33

Summaries of the subject, 1-2, 1-3

Surf noise, 3-17, 7-18

Surface duct, 4-12, 6-14

*****

Thermal noise, 2-27

Time-delay correlograms, 6-10

Trident Vertical Array (TVA), 5-6, 6-2, 6-8

Twenty-cycle pulses, 7-11

*****

Variability of noise:
  averaging time, 3-1
  beamwidth, 3-18
  deep water, 3-3
  detectability, 3-21
  frequency bands, 3-18
  narrow bands, 3-4
  shallow water, 3-12

Velocity profile, effects on noise, 4-14

*****

Wave interaction, as a noise source, 2-32

Whales, dolphins and porpoises, 7-2, 7-12

Wind noise:
  level of, 2-23
  processes of generation, 2-30

WOTAN, 2-27

Synopses of *Peninsula Publishing* Books

Sound Propagation in the Sea — the most up-to-date text on underwater sound propagation in today's literature. The author of *Sound Propagation* is Robert J. Urick, leading lecturer and author of *"Principles of Underwater Sound,"* a classic in the field. Originally written for the Defense Advanced Research Projects Agency, this revised edition incorporates corrections and an additional chapter on underwater propagation modelling. Subjects include: Basic Theory; Velocity of Sound in the Ocean; Attentuation and Absorption; The Surfact Duct; The Deep Sound Channel; Caustics and Convergence Zones; Computer Models of Sound Propagation; Shallow-Water Ducts; Reflections and Scattering by the Sea Surface; Reflections and Scattering by the Sea Bottom; Temporal Coherence; Spatial Coherence; and Multipath in the Sea.

Space-Time Information Processing — A classic. Written by Charles Loda and A. Winder. Indispensable reference tool for audio signal processing and data analysis. Oriented to acoustics and sonar engineering. The book features Fourier transforms, statistical analyses, correlation and spectra. Especially valuable chapters address measurement of spatially and temporally-limited functions, optimal filtering procedures and interpretation of results.

Microwave Receivers and Related Components — Dr. James Tsui of the USAF Avionics Laboratory. Hardcover, 530-pages. The best one-volume reference on receivers in print today. **Receivers:** Seven chapters devoted to receiver technology: crystal video and superheterodyne receivers, IFM, channelized, compressive and Bragg cell receivers. Chapters discuss characteristics, receiver capabilities and limitations, basic design and principles of operation, mathematical analysis, probability of intercept, applications, and receiver evaluation. **Components:** Another eight chapters provide a systems overview of the fundamental components of microwave receivers: transmission lines; delay lines; directional couplers; hybrid couplers and power dividers; attenuators, circulators and phase shifters; filters; linear and logarithmic amplifiers and limiters; mixers and detectors; oscillators and frequency synthesizers.

Microwave Scanning Antennas — R. C. Hansen. Three-volumes bound into a single hardcover book. 1200 pages. This classic reference covers all types of narrow beam antennas with emphasis on those that scan the beam. Comprehensively treats phased arrays and quasi-optical antennas. Includes multiple-beam arrays, time domain and synthetic apertures and adaptive antennas. Concentrates on narrow beam antennas, both mechanically steered and electronically scanned. Constraining equations and design tradeoffs along with practical examples. Volume I reviews on continuous aperture, theory and practice, optical scanners, and two closely related areas, radio astronomy instruments and large radomes. Volumes II and III cover array theory; a variety of means for scanning such as phase shift, frequency and heterodyne; and the systems technology of time domain and adaptive antenna systems.

Principles of High-Resolution Radar — This 500 page book written by August Rihaczek features the finest 3-dimensional computer derived depictions of the ambiguity function published anywhere! The book presents a step-by-step development of the theory of radar resolution beginning with basic measurements on single targets and proceeds through an analysis of the nature of the resolution problem. Capabilities and limits of radar, and the details of waveform design are discussed. Synthetic aperture radar is used to illustrate the practical application of principles, and derivations are carefully spelled out and supported by extensive reasoning and interpretation. Principles are stressed as the author clearly conveys the theory of target resolution without resorting to advanced mathematics. Although geared primarily to the practicing radar engineer and technical specialists, this volume will be useful to the non-specialist and student as well. Topics covered include: measurements, ambiguities, and pulse compression for single targets; target resolution and the limits on resolution performance; target detection in various forms of clutter and the relation to resolution; waveform design; higher order range derivatives, Doppler distortions, wideband signals, and general types of target motion.

The Strategy of Electromagnetic Conflict — Written by the Air Force Academy faculty. Contents include: Electronic Countermeasures and Counter-countermeasures; electronic reconnaissance; tradeoffs in air defense; use of chaff for aircraft self-protection; communications intelligence, security and cryptology; vulnerability of space stations; countermeasures in the millimeter, IR, optical, laser and UV regions. *Strategy's* 300 pages are attractively packaged in a hardbound edition. Required reading for engineering and analysis dealing with design and development of military electronic systems and the planning of air operations.

Electronic Countermeasures — Originally published as a Secret reference in the 1960's by the U.S. Army Signal Corps, the book is now declassified. Its 1100 pages cover Signal Intercept, Jamming and deception fundamentals that are as valid today as when first written. Other subjects include intercept probabilities, receiver parameters, detection and analysis, direction finding, jamming technique, IR and acoustic countermeasures. More than 600 references and a list of authors that reads like Who's Who in Electronic Warfare.

Principles of Electronic Warefare — This reprinted version provides an overview of EW technology and tactics in the space age. Written by Robert J. Schlesinger from and overall systems viewpoint, the book focuses on EW's contribution in the accomplishment of a military aircraft mission. Some highlights: radar countermeasure techniques; electronic intelligence (ELINT); elements of space environment; payload optimization between weapons and ECM; and air combat analysis.

Digital Communications with Space Applications — this book defined an entire new technology for space communications. It was known, irreverantly, as "The Bible" at the Jet Propulsion Laboratory. Subjects addressed include: C^3, data telemetry, tracking and ranging, coding, sequences and synchronization techniques. The authors, Solomon W. Golomb, Andrew J. Viterbi, Leonard Baumert, Mahlon Easterling and Jack Stiffer are among the foremost experts in the communications field.

Methods of Operation Research — One of the first books exclusively devoted to Operations Research (OR). Kimball and Morse examine the application of OR to decision-making problems in military operations. Topics include: probability; measures of effectiveness; tactical analyses; Lanchester equations; evaluation of equipment performance; mini-max; exchange rate; sweep rates; operations involving men and equipment; search theory; plus more. A highly articulate work that has special value to Systems Engineers and Operations Researchers.

Proposal Management — the Modular Technique can help you achieve effective proposal results! It establishes a vehicle for the communication of complex ideas between manager and authors. It facilitates a large group of authors documenting ideas in a cohesive, easily-read proposal. The *Modular Technique,* based on storyboarding procedures as outlined in this book, provides the manager with the technique for planning the work, assigning design and writing tasks, implementing author-to-manager feedback and effecting document review — in a timely manner. If you are involved with proposals, this book will help you prepare winning proposals in a disciplined, orderly and cost-effective fashion. Indeed, the entire proposal team will give you a loud hurrah!

Postcard to order books on next page